New Perspectives in Behavioral Cybersecurity

New Perspectives in Behavioral Cybersecurity offers direction for readers in areas related to human behavior and cybersecurity by exploring some of the new ideas and approaches in this subject, specifically with new techniques in this field coming from scholars with very diverse backgrounds in dealing with these issues. It seeks to show an understanding of motivation, personality, and other behavioral approaches to understand cyberattacks and create cyber defenses.

This book:

- Elaborates cybersecurity concerns in the work environment and cybersecurity threats to individuals.
- Presents personality characteristics of cybersecurity attackers, cybersecurity behavior, and behavioral interventions.
- Highlights the applications of behavioral economics to cybersecurity.
- Captures the management and security of financial data through integrated software solutions.
- Examines the importance of studying fake news proliferation by detecting coordinated inauthentic behavior.

This title is an ideal read for senior undergraduates, graduate students, and professionals in fields including ergonomics, human factors, human-computer interaction, computer engineering, and psychology.

Front Cover

New Perspectives in Behavioral Cybersecurity

Photographs:

Top Row, L to R: Palvi Aggarwal, University of Texas–El Paso, USA; Gerardo Armenta, University of Texas–El Paso, USA; Jeremy Blackstone, Howard University, USA; Rodney H. Cooper, University of New Brunswick, Canada; Rozalina Dimova, Technical University of Varna, Bulgaria.

Second Row, L to R: Tihomir Dovramadjiev, Technical University of Varna, Bulgaria; Rusko Filchev, Technical University of Varna, Bulgaria; Marton Gergely, United Arab Emirates University, UAE; Kristjan Karmo, Tallinn University of Technology, Estonia; Kaido Kikkas, Tallinn University of Technology, Estonia.

Third Row, L to R: Birgy Lorentz, Tallinn University of Technology, Estonia; Sten Mäses, Tallinn University of Technology, Estonia; Thomas Morris and Jeremiah Still, Old Dominion University, USA; Ebelechukwu Nwafor, Villanova University, USA; Augustine Orgah, Xavier University of New Orleans, USA.

Fourth Row, L to R: Wayne Patterson, Patterson and Associates, USA and Canada; Heba Saleous, United Arab Emirates University, UAE; Brandon Sloane, New York University, USA; Michael Sone, University of Buea, Cameroon; Ange Taffo, University of Buea, Cameroon.

Bottom Row, L to R: Gloria Washington, Howard University, USA; William Yu, Ateneo de Manila University, Philippines.

New Perspectives in Behavioral Cybersecurity

Human Behavior and Decision-Making Models

Edited by
Wayne Patterson

CRC Press is an imprint of the
Taylor & Francis Group, an informa business

Designed cover image: The front cover demonstrates the highly diverse nature of the authors, who are also from ten countries on five continents.

First edition published 2024
by CRC Press
2385 NW Executive Center Drive, Suite 320, Boca Raton FL 33431

and by CRC Press
4 Park Square, Milton Park, Abingdon, Oxon, OX14 4RN

CRC Press is an imprint of Taylor & Francis Group, LLC

Library of Congress Cataloging-in-Publication Data
Names: Patterson, Wayne, 1945– editor.
Title: New perspectives in behavioral cybersecurity : human behavior and decision-making models / edited by Wayne Patterson.
Description: First edition. | Boca Raton : CRC Press, 2024. |
Includes bibliographical references and index.
Identifiers: LCCN 2023009969 (print) | LCCN 2023009970 (ebook) |
ISBN 9781032414775 (hbk) | ISBN 9781032540818 (pbk) | ISBN 9781003415060 (ebk)
Subjects: LCSH: Behavioral cybersecurity. | Internet users–Psychology. |
Computer networks–Security measures.
Classification: LCC ZA3086.5.C93 N49 2024 (print) | LCC ZA3086.5.C93 (ebook) |
DDC 303.48/34–dc23/eng/20230717
LC record available at https://lccn.loc.gov/2023009969
LC ebook record available at https://lccn.loc.gov/2023009970

ISBN: 9781032414775 (hbk)
ISBN: 9781032540818 (pbk)
ISBN: 9781003415060 (ebk)

DOI: 10.1201/9781003415060

Typeset in Times
by Newgen Publishing UK

Contents

Preface

NEW PERSPECTIVES IN BEHAVIORAL CYBERSECURITY

Since the introduction and proliferation of the Internet, problems involved with maintaining cybersecurity have grown exponentially and evolved into many forms of exploitation.

Yet, cybersecurity has had far too little study and research. Virtually all of the research that has taken place in cybersecurity over many years has been done by those with computer science, electrical engineering, and mathematics backgrounds.

However, many cybersecurity researchers have come to realize that to gain a full understanding of how to protect a cyber environment requires not only the knowledge of those researchers in computer science, electrical engineering, and mathematics, but those who have a deeper understanding of human behavior: researchers with expertise in the various branches of behavioral science, such as psychology, behavioral economics, and other aspects of brain science.

The contributors, arising from several disciplines, have attempted in the past few years to understand the contributions that distinct approaches to cybersecurity problems can benefit from; this interdisciplinary approach is what we have tended to call "behavioral cybersecurity."

Our decision to create this book arose from a similar perspective. Our training is in computer science and psychology, among other fields, and we have observed, as have many other scholars interested in cybersecurity, that the problems we try to study in cybersecurity require not only most of the approaches in computer science, but more and more an understanding of motivation, personality, and other behavioral approaches in order to understand cyberattacks and create cyber defenses.

NEW SCHOLARS IN BEHAVIORAL CYBERSECURITY

As with any new approaches to solving problems when they require knowledge and practice from distinct research fields, there are few people with knowledge of the widely separate disciplines, so it requires an opportunity for persons interested in either field to gain some knowledge of the other. We have attempted to provide such a bridge in this book that we have entitled *New Perspectives in Behavioral Cybersecurity*.

In this book, we have tried to provide an introductory approach to both psychology and cybersecurity, and as we have tried to address some of these key problem areas, we have also introduced topics from other related fields such as criminal justice, game theory, mathematics, and behavioral economics.

We entered the computer era almost 75 years ago. For close to two-thirds of that time, we could largely ignore the threats that we now refer to as cyberattacks. There were many reasons for this. There was considerable research done going back to the 1970s about approaches to penetrate computer environments, but there were several other factors that prevented the widespread development of cyberattacks. Thus, the scholarship into the defense (and attack) of computing environments remained of interest to a relatively small number of researchers, primarily from mathematics and computer science.

Beginning in the 1980s, a number of new factors came into play. First among these was the development of the personal computer, which now allowed for many millions of new users with their own individual access to computing power. Following closely on that development was the expansion of network computing, originally through the defense-supported DARPAnet, which then evolved into the openly available Internet. Now, and with the development of tools such as browsers to make the Internet far more useful to the world's community, the environment was set for the rapid expansion of cyberattacks, both in number and in kind, so the challenge for cybersecurity researchers over a very short period of time became a major concern to the computing industry.

The world of computer science was thus faced with the dilemma of having to adapt to changing levels of expertise in a very short period of time. The editor of this book began his own research in 1980, in the infancy of what we now call cybersecurity, even before the widespread development of the personal computer and the Internet.

CONTRIBUTIONS FROM DIVERSE DISCIPLINES

In this book, we have tried to achieve several objectives in addressing a number of issues involving cybersecurity threats in many different types of environment. In addition, we have come to understand that the entire cyber environment in which most of us live and participate, has threats that can emerge from many directions and at many levels of sophistication.

As interest in a broad range of approaches to provide greater opportunities for defending and securing one's computer environment has grown, we find that there is a rapidly growing field of scholars, emerging both from fields of computer science and cybersecurity, and also from a range of behavioral sciences including psychology, criminal justice, law, sociology, and other human behavior-related disciplines.

In addition, we have noted and present in this book, contributions that illustrate that this aspect of cybersecurity, once thought to be only studied in a few highly developed countries, now has serious scholarship in a wide range of countries all over the world. This book features contributions from, for example, Bulgaria, Cameroon, Canada, Estonia, the Philippines, and United Arab Emirates as well as the United States.

Scholarship in this field has interested many new young researchers, as indicated by many of the contributors to this book, not only coming from diverse national environments, but also illustrating that this aspect of cybersecurity has been an interest in encouraging scholarship from a significant number of women scholars—close to 30% of our contributing authors. This is notable as over the last three decades of scholarship when both computer science and the field of cybersecurity began to evolve, the topic of cybersecurity had very few contributions from over 30% of the population, very few of these from women scholars.

Of the research articles presented in this book, we have somewhat arbitrarily divided them into five groups, as follows:

1. Cybersecurity Concerns in the Work Environment

We begin with a contribution from Rusko Filchev, Tihomir Dovramadjiev, and Rozalina Dimova of the Technical University of Varna in Bulgaria, addressing the subject of "Management and Security of Financial Data Through Integrated Software Solutions." Continuing to address issues in the workplace, we include a contribution from Michael Sone and Ange Taffo of the University of Buea in Cameroon, "An Efficient Scheme for Detecting and Mitigating Insider Threats." Finally, for this section, Gerardo Armenta and Palvi Aggarwal of the University of Texas at El Paso, add "Phishing Through URLs", based on an approach to detecting the extent to which people are likely to be fooled by phishing URLs.

2. Cybersecurity Threats to the Individual

Given a number of different environments that the casual user may explore, we address questions involving threats from video games, from ransomware or other scams, and the confusion spread by quote "fake news."

The first contribution in this section comes from Heba Saleous and Marton Gergely of the United Arab Emirates University, addressing video games environments, since we have learned that cybercriminals have found that these games provide more and malicious opportunities. Professor Birgy Lorenz and her colleagues, Sten Mäses, Kaido Kikkas, and Kristjan Karmo of the Tallinn University of Technology in Estonia, contribute "Dances with the Illuminati", a methodology for replying to phishing scams as a mechanism for learning about the criminal modes of operandi. For the third approach in dealing with Internet threats, William Yu of the Ateneo de Manila University in the Philippines, has written "Studying Fake News Proliferation by Detecting Coordinated Inauthentic Behavior", in illustrating a mechanism to catch purveyors of disinformation by studying their network relationships.

Rodney Cooper of the University of New Brunswick, together with editor Wayne Patterson, contribute an article demonstrating a method to determine if a text originated in a different language and had been translated: "Refining the Sweeney Approach on Data Privacy."

3. Cybersecurity Concerns in the Home and Work Environment

Thomas Morris and Jeremiah Still of the Old Dominion University in Norfolk VA, have contributed "Cybersecurity Hygiene: Blending Home and Work Computing." And they note that poor cybersecurity behaviors can cost companies and individuals millions of dollars, citing as an example the 2021 ransomware attack of a US oil pipeline system; the hacker's demand was for $4.4 million to unencrypt pipeline operations. Following this, Augustine Orgah of Xavier University of New Orleans poses the question, certainly related to the previous contribution, "Will a Cyber Security Mindset Shift, Build, and Sustain a Cyber Security Pipeline?"

4. Ethical Behavior

Brandon Sloane of the New York University contributes "Cybersecurity Behavior and Behavioral Interventions." His article examines ethical decision-making models and synthesizes them to define a high-level decision-making process.

5. Differences in Language in Cyberattacks

It is the case that in most cyberattacks, or messages to be read by these victims, the issue of expression of a demand or attack must be presented in a specific human language for the potential victim to understand.

In terms of a cyberattacker from a different part of the world with a different language base, one means of detecting an attack text is to attempt to discover if the attack message has originated in a different language and has been translated. In "Using Language Differences to Detect Cyberattacks: Ukranian and Russian", Wayne Patterson examines, following several other studies, what might be appropriate to the recent Russia/Ukraine conflict by comparing texts in the Russian and Ukrainian languages.

6. Applications of Behavioral Economics to Cybersecurity

Jeremy Blackstone of Howard University addresses an important branch of research, which has emerged from what is known as behavioral economics, in his paper "Using Economic Prospect Theory to Quantify Side Channel Attacks."

Ebelechukwu Nwafor of Villlanova University in Philadelphia utilizes principles of game theory in "A Game-Theoretic Approach to Detecting Romance Scams."

7. New Approaches for Future Research

In this section, Gloria Washington of Howard University, addresses the security issues in the use of motion detection in the creation of non-verbal password mechanisms in her contribution "Human-Centered Artificial Intelligence: Threats and Opportunities for Cybersecurity."

FUTURE DIRECTIONS IN BEHAVIORAL CYBERSECURITY

It is believed that in order to counter the clever but malicious behavior of hackers and the sloppy behavior of honest users, cybersecurity professionals (and students) must gain some understanding of motivation, personality, behavior, and other theories that are studied primarily in psychology and other behavioral sciences.

Consequently, by building a behavioral component into a cybersecurity program, it is felt that this curricular need can be addressed. In addition, noting that while only 20% of computer science majors in the United States are women, about 80% of psychology majors are women. It is hoped that this new curriculum, with a behavioral science orientation in the now-popular field of cybersecurity, will induce more women to want to choose this curricular option.

In terms of employment needs in cybersecurity, estimates indicate "more than 209,000 cybersecurity jobs in the US are unfilled, and postings are up 74% over the past five years."

It is believed that the concentration in behavioral cybersecurity will also attract more women students since national statistics show that whereas women are outnumbered by men by approximately 4 to 1 in computer science, almost the reverse is true in psychology.

It has also not escaped our notice that the field of cybersecurity has been less attractive to women. Estimates have shown that even though women are underrepresented in computer science (nationally around 25%), in the computer science specialization of cybersecurity, the participation of women drops to about 10%.

However, with the development of a new path through the behavioral sciences into cybersecurity, we observed that approximately 80% of psychology majors, for example, are female. We hope that this entrée to cybersecurity will encourage more behavioral science students to choose this path, and that computer science, mathematics, and electrical engineering students interested in this area will be more inclined to gain a background in psychology and the behavioral sciences.

Acknowledgments

The editor (Patterson) is extremely grateful for the learned contributions from the 22 coauthors who have provided important parts of this volume.

Editor Biography

Dr. Wayne Patterson is a retired Professor of Computer Science at Howard University. He is also currently Co-Principal Investigator for the GEAR UP project at Howard. He has also been Director of the Cybersecurity Research Center, Associate Vice Provost for Research, and Senior Fellow for Research and International Affairs in the Graduate School at Howard. He has also been Professeur d'Informatique at the Université de Moncton, Chair of the Department of Computer Science at the University of New Orleans, and in 1988, Associate Vice Chancellor for Research there. In 1993, he was appointed Vice President for Research and Professional and Community Services, and Dean of the Graduate School at the College of Charleston, South Carolina. In 1998, he was selected by the Council of Graduate Schools, the national organization of graduate deans and graduate schools, as the Dean in Residence at the national office in Washington, DC. His other services to the graduate community in the United States have included being elected to the Presidency of the Conference of Southern Graduate Schools and also to the Board of Directors of the Council of Graduate Schools. Dr. Patterson has published more than 50 scholarly articles primarily related to cybersecurity.

He also published one of the earliest textbooks in cybersecurity, *Mathematical Cryptology* (Rowman and Littlefield, 1986), and recently the first book in this aspect of cybersecurity (with coauthor Cynthia K. Winston), *Behavioral Cybersecurity* (CRC Press, 2018). He has been the principal investigator on over 35 external grants valued at over $6,000,000. In August 2006, he was loaned by Howard University to the US National Science Foundation to serve as the Foundation's Program Manager for International Science and Engineering in Developing Countries, and in 2017 was Visiting Scholar at Google.

He received degrees from the University of Toronto (BSc and MSc in Mathematics), the University of New Brunswick (MSc in Computer Science), and the University of Michigan (PhD in Mathematics). He also held post-Doctoral appointments at Princeton University and the University of California–Berkeley.

Editor Biography

[illegible]

[illegible]

[illegible]

[illegible]

Contributors

Aggarwal, Palvi
University of Texas–El Paso, USA

Armenta, Gerardo I.
University of Texas–El Paso, USA

Blackstone, Jeremy
Howard University, USA

Cooper, Rodney H.
University of New Brunswick, Canada

Dimova, Rozalina
Technical University of Varna, Bulgaria

Dovramadjiev, Tihomir
Technical University of Varna, Bulgaria

Filchev, Rusko
Technical University of Varna, Bulgaria

Gergely, Marton
United Arab Emirates University, UAE

Karmo, Kristjan
Tallinn University of Technology, Estonia

Kikkas, Kaido
Tallinn University of Technology, Estonia

Lorenz, Birgy
Tallinn University of Technology, Estonia

Mäses, Sten
Tallinn University of Technology, Estonia

Morris, Thomas
Old Dominion University, USA

Nwafor, Ebelechukwu
Villanova University, USA

Orgah, Augustine
Xavier University of New Orleans, USA

Patterson, Wayne
Patterson and Associates, USA

Saleous, Heba
United Arab Emirates University, UAE

Sloane, Brandon
New York University, USA

Sone, Michael E.
University of Buea, Cameroon

Still, Jeremiah
Old Dominion University, USA

Taffo, Ange
University of Buea, Cameroon

Washington, Gloria
Howard University, USA

Yu, William Emmanuel S.
Ateneo de Manila University, Philippines

Section I

Cybersecurity Concerns in the Work Environment

1 Management and Security of Financial Data Through Integrated Software Solutions

Rusko Filchev, Tihomir Dovramadjiev and Rozalina Dimova
Technical University of Varna
Varna, Bulgaria

1 INTRODUCTION

The current stage of the formation of the information society (local and global) is characterized by the large-scale introduction of information and communication technologies (ICT) in all important areas of life activity (Arellano & Cámara, 2017 and Haddon, 2004). Information resources, the emergence and development of new forms and types of virtualization and e-interaction (e-government, e-government and training, e-commerce, Internet and mobile banking, telemedicine, social networks and e-publications, etc.) are an integral part of society (Polyakov, 2016, Roztocki et al. 2019 and Aksentijević et al., 2021).

Managing and securing financial data is a major global challenge. This activity that accompanies people's lives is an integral part of practically every person. Whether in a personal, corporate, or public plan the standard procedures for working with finances are performed, they are in one way or another managed by the person/s taking into account the conditions in which they are. Whether in

DOI: 10.1201/9781003415060-2

private or business, literacy in the proper handling of financial data is a matter of paramount importance.

A serious aspect is the security of financial data, as well as access control. It is desirable to be able to create accounts protected by passwords. Financial management software (FMS) is a tool for regulating financial operations by illustrating the actions performed covering periods and regulating the overall balance of users (Savina & Kuzmina-Merlinob, 2015, Hashim, 2014 and USAID, 2008). The ability to handle financial data is an important condition for proper and sustainable budget management. When working with financial transactions, it is necessary to take into account all aspects of personal data protection, as well as ensuring technical security and in accordance with legal provisions (HD, 2021, De Jesus, 2004 and Feyen et al. 2021).

1.1 Security of Personal Data

Respect for the right to privacy and the right to the protection of personal data are a priority for democratic societies. Technology has its place and certain benefits for people and society, in some cases improving quality of life, efficiency, and productivity. At the same time, they create risks to the right to respect for private life. In response to the need for special rules governing the collection and use of personal information, there is also a need to create a concept of privacy, known in some jurisdictions as "privacy of personal information" and in others as "the right to information self-determination" (EUAFRCE, 2019).

1.2 Occurrence of Electronic Databases

Historically, electronic database arrays were created in the 1960s (Schneier, 2000). This was the first stage of large-scale electronic storage. Computers allowed the accumulation of relatively large databases. Public organizations began to accumulate databases containing information about individuals. Subsequently, the second critical limit was reached: computer networks allowed sharing. This was then used for sharing, comparing, and merging separate databases. In the following years, systems began to collect and store more data due to the fact that the collection of information became more accessible and cheaper, as well as due to the fact that people became more electronic fingerprinted–"trace"–in everyday life such as to operate electronic devices. Due to globalization and the widespread use of online networks, a significant part of the data in one way or another is collected, stored, compared, etc. People start building an online file consciously or unconsciously depending on their activities. The online network offers more opportunities to violate privacy. In theory, e-shops can keep records of everything that is bought. For public organizations, for example, operational databases are a huge help to law enforcement police. They help to automatically receive information and photos directly to the patrol car, but the risk of confidentiality remains. This shows that such socially important databases as the police database need good control and confidentiality (Schneier, 2000).

1.3 Need and Security of Electronic Data

The global world is in a stage of rapid technological development and constant transformations are taking place in all sectors of the world economy and the system of corporate governance, and therefore the system of corporate security cannot work according to the old standards (Zakharov, 2021). These requirements can also be considered personally, where personally or organizationally it is necessary to optimally adapt to modern conditions. The need leads to greater requirements. Implementation of modern trends in the optimization of business processes related to the work with electronic data based on digitalization technologies provides great opportunities. At the same time, the processes of ensuring the security of electronic data are significantly complicated.

People have different understandings of need and security. Also, this is accompanied by the specific need. When dealing with electronic data related to financial and other transactions in the online space, a process of chaotic actions is often observed. Given the variety of activities performed, they are performed in different places, which in turn create certain random actions that directly contradict the synchronized and well-planned budgeting. These features are also observed in certain commercial companies and other companies not well acquainted or unprepared for well-planned digital consumption and/or not implemented modern budgeting software (Karlson, 2019).

Public organizations are more purposefully structured in terms of their direct activities. This is also directly related to security. As a rule, public organizations pay more attention to their security than individuals. Organizations must also communicate to people the reasons for collecting information, provide access to it, correct inaccuracies, and protect this information from unauthorized parties. People have the right to see the personal data collected about them and to correct the errors in them. They also have the right to know why this information is being collected and to ensure that the information is not sold for other purposes. And they also have the right to "avoid" any intelligence gathering when they don't want to. Data collectors should be responsible for the protection of individual data to a relatively high degree and should not share data with anyone who does not adhere to these rules (Schneier, 2000).

The Law on Personal Data Protection is valid in Bulgaria (State Gazette No. 7 of 19 January 2018–active) (CMRB, 2018).

1.4 Objects of Informational Security

The objects of information security in companies and organizations include (Yasenev et al., 2017):

- Informational resources. They contain information classified as trade secret and confidential information presented in the form of information files and a database;
- Tools and system of informatization. Includes computing devices, organizational technologies, networks and systems, system-wide and application software, automated control systems, communication and data transmission systems, technical means for collecting, registering, transferring, processing, and displaying information.

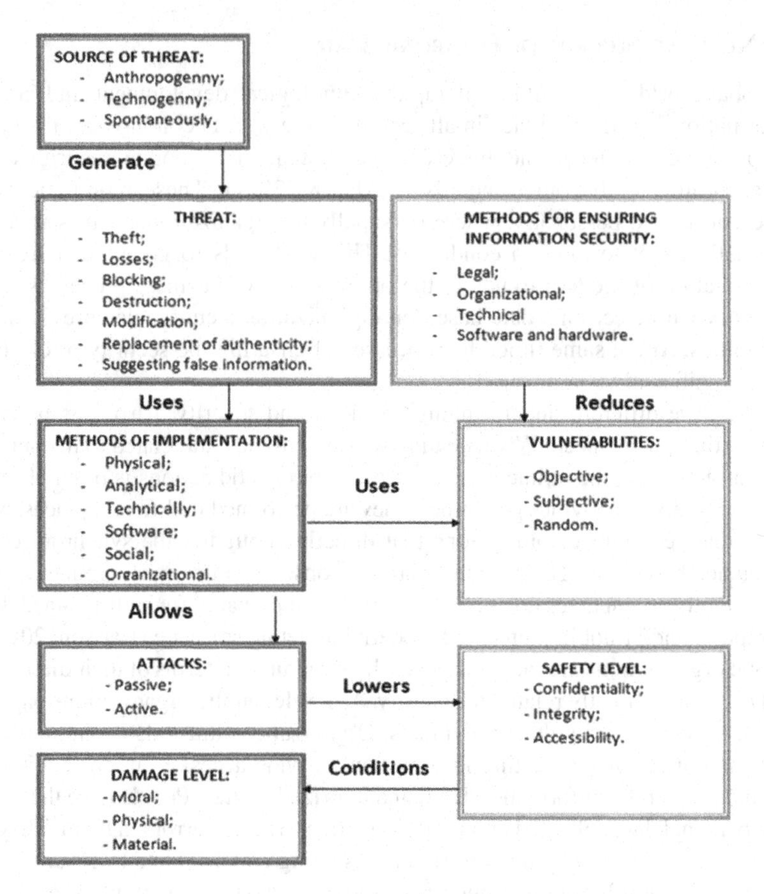

FIGURE 1.1 Information security threat model (Azhmukhamedov, 2012).

1.5 Information Security Threat

The threats to information security are presented in Figure 1.1 (Azhmukhamedov, 2012).

1.6 Information Security Management and Audit

Providing one's own information security of the company or a public organization is a necessity (Nieles et al., 2017). When it comes to financial data transactions, it means working with an array of information that is related to capital. This makes the activity large even in the context of the individual without significant scale and value of money in relation to the individual financial micro or macro climate. Referring to companies, organizations, etc., then, as a rule, the provision of information security is an integral part of the overall management system (Sahar et al., 2017). It is necessary to achieve statutory goals and objectives. The importance of systematic targeted activities to ensure information security becomes greater, the greater the degree of automation of business processes of the company and/or public organization, and the greater the "intellectual component" in its final product, i.e. the success of the activity depends very much on the availability and preservation of certain information, ensuring its confidentiality and accessibility for owners and consumers (Yasenev et al., 2017).

In addition to the state level, information security management at the enterprise level is aimed at neutralizing various types of threats:

- External. They refer to illegal actions of state bodies, illegal actions of criminals and criminal groups, illegal actions of competing companies and other economic entities, unfair actions of partner companies, inconsistency of the current regulatory framework with the real development of technologies, and public relations, failures, and disturbances in the operation of global information and telecommunication systems, etc;
- Internal. They refer to personal mistakes and negligence of the personnel of the enterprise, as well as intentionally recognized violations, failures, and irregularities in their own work, information systems, etc.

Thus, the management of information security in each individual enterprise must be carried out in the context of its overall economic activity: taking into account the nature of the company's activity, as well as the real situation in market competition, government policy, and development of legal and enforcement systems. The level of development of some used information and telecommunication technologies and other factors that form the general conditions of the current activities must be assessed.

In order to neutralize the existing threats and to ensure information security, the companies, as well as the public organizations, establish a management system in the field of information security, within which (the systems) work in several areas (Yasenev et al., 2017):

- Formation and practical implementation of a comprehensive multi-level information security policy and a system of internal requirements, norms, and rules (John & Hoffman, 2019);
- Organization of the service department for information security (Hammond & Gummer, 2016);
- Development of a system of measures and actions in case of unforeseen situations ("Incident Management") (Barrington, 2019);
- Carrying out audits (comprehensive checks) of the state of information security in the enterprise (Zegjda & Ivashko, 2005).

The structure of the organizational activities in the field of information security is shown in Figure 1.2 (Yasenev et al., 2017).

1.7 Security of Personal Data

The presence of ICT and modern automation technologies increase the amount of information processed electronically, which leads to a reduction in the overall level of security in the work of the bank (Cetin Karabat & Karabat, 2012). The solution to

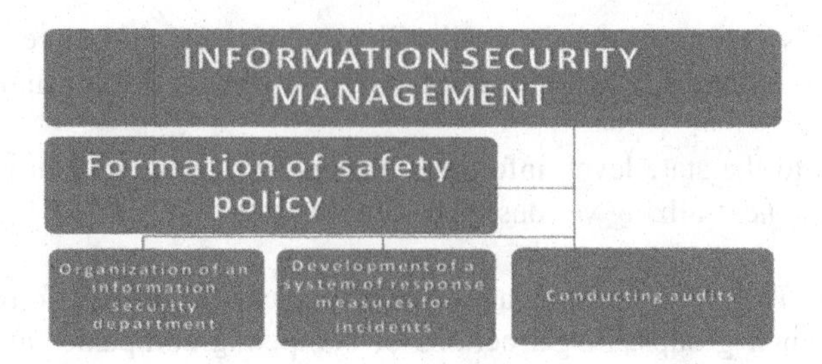

FIGURE 1.2 Structure of organizational activities in the field of information security (Yasenev et al., 2017).

this problem largely depends on the technology used by a particular bank, i.e. from the automated banking system. The computerization of banking activities makes it possible to significantly increase the productivity of bank employees, and the introduction of new financial products and technologies (Tuli & Juneja, 2016). At the same time, criminal activity using certain technologies is also becoming more dangerous for the security of financial data (Jang-Jaccard & Nepal, 2014). Currently, over 90% of all crimes are related to the use of automated information processing systems of banks. When creating and modernizing the automation of banking systems (ABS) for working with financial data, it is necessary to pay great attention to ensuring security. This problem needs in-depth study. If established physical approaches have long been developed to ensure physical and classical information security, then due to the frequent radical changes in computer technology, the methods of ABS protection require constant updating (Yasenev et al., 2017).

2 FINANCIAL DATA MANAGEMENT

2.1 Modeling of Information Culture

Working with financial data requires certain skills from the manager who works with and manages them. Regardless of whether he is a specialist or not, the initial modeling of information culture (IC) is necessary. IC is a complex, multidimensional, continuous step-by-step process of creating information skills based on professional and/or personal training. The level of IC is an indicator of the individual's readiness for self-improvement and self-development. The level of IC is characterized by the efficiency of use of information resources, adaptability to educational and professional activities in the information space, formed on the basis of modern ICT with mandatory provision of information security requirements (Figure 1.3) (Polyakov, 2016).

2.2 Realization of the Financial Balance

Achieving financial balance (FB) through good management is one of the most important things that is regularly done by a person, family, company, organization, etc. The main goal of optimized budgeting is to perform analysis, control, and timely

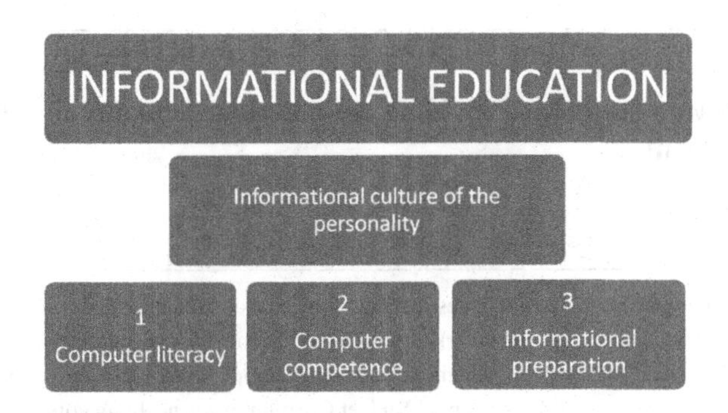

FIGURE 1.3 Gradual formation of the information culture on the basis of information and communication technologies (ICT) (Polyakov, 2016).

identification and elimination of shortcomings in financial activities. Reliable balance sheet data is needed by owners to control available capital. This is of great importance for the planning processes (Kozhin et al., 2016) and (Morgunova, 2020).

The FB contains a large amount of data on the activities of the user(s). The FB includes [30]:

- General financial condition;
- Characteristics of the state of resources;
- Availability of funds;
- Investments.

2.3 Budget Planning

Every adult member of society is responsible for treating the distribution of their own and/or family budgets responsibly (Kozhin et al., 2016) and (Morgunova, 2020). Referring to business and public organizations, budget planning (budgeting) becomes possible taking into account certain conditions, where the implementation contains the following aspects and features: budget or forecast planning (Shakhovskaya et al., 2009), a quantitative plan in monetary terms and information system for domestic production management, partial costs (Egorshin, 2009), and integrated planning–end-to-end system process of preparation and implementation of the budget in practical activities (Shchiborsch, 2001).

3 INTEGRATED SOFTWARE SOLUTIONS

For the needs of financial operations (budgeting), it is necessary to implement integrated software solutions. These are technical software tools through which certain operations for entering financial data, values, available resources, etc., are performed. Effective software solutions providing the necessary quality for financial data management are shown in Table 1.1. Given the diverse activities of users, it is necessary to provide such software solutions covering the maximum range of capabilities for

TABLE 1.1
Effective Software Solutions Providing the Necessary Quality for Financial Data Management

Integrated Budgeting Software	Software/Homepage Link
Budgeting software in public systems (Adaptive Planning, 2022 and G2, 2022)	• Questica. www.questica.com/budget/ • Workday Adaptive Planning. www.adaptiveplanning.com/uk/solutions/budgeting-and-planning-software-for-the-public-sector • Capital Budget Creation. www.neubrain.com/capital-budget-software-government
Budgeting software designed for business-oriented companies and commercial companies (Karlson, 2022)	• Centage. www.centage.com/ • Prophix. www.prophix.com/ • Float. https://floatapp.com/ • Planguru. www.planguru.com/ • GIDE. https://mygide.com/ • Tagetik. www.wolterskluwer.com/en/solutions/cch-tagetik • Adaptive Insights. www.adaptiveplanning.com/ • Coupa. www.coupa.com/products/procurement/budgets • Tidemark. https://insightsoftware.com/tidemark/ • Neubrain. www.neubrain.com/ • QuickBooks. https://quickbooks.intuit.com/ • Other
Personal finance budgeting software (Cox et al., 2021)	• Quicken. www.quicken.com • YNAB. www.youneedabudget.com/ • Banktree. www.banktree.co.uk/ • Other
Free & Open budgeting software (L.F., 2022 and Whitt, 2015)	• GnuCASH. https://gnucash.org/ • Money Manager Ex. www.moneymanagerex.org/ • HomeBank. http://homebank.free.fr • AbilityCash. https://dervish.ru/ • Other

operating data, storing them in file formats, as well as creating personal or group accounts with the necessary levels of access and passwords.

- Software for budgeting in public systems. These are public systems—financial, health, social, judicial, etc. As a rule, these are software that are specially designed to serve these areas and work on their own, and are sometimes integrated with other systems of other social systems depending on how this process is regulated (Adaptive Planning, 2022) and (G2, 2022);
- Budgeting software designed for business-oriented companies and commercial companies. They are specially created for the needs of the respective company or are commercially insured—licenses are purchased, etc. Software that is designed to serve a wide range of users (Karlson, 2022);

- Personal finance budgeting software. This group is of great interest because this type of system can be applied by one person and extended to multiple users, making them multifunctional and well applicable in a number of areas (Cox et al., 2021);
- Free & open source budgeting software. Modern open source budgeting software is in constant technological progress and is fully functionally efficient (L.F., 2022 and Whitt, 2015).

It is important that when choosing budgeting software, it has an appropriate set of both functionality and storage of the entered data in file formats that can be transferred to other software and platforms when needed. Attention should also be paid to data protection by encrypting and creating passwords. This will provide an opportunity for preventative protection against potential malicious actions.

4 APPLICATION OF FREE & OPEN SOURCE BUDGETING SOFTWARE AND EXPERIMENTS ON FINANCIAL DATA TRANSFER BETWEEN DIFFERENT SYSTEMS AND PROVIDING SECURITY BY ENCRYPTION

4.1 Main Features

Open source budgeting software is preferred mainly because it is easy to access and most of them run on different OSes, so also have installed and/or portable versions. Leading OS software such as GnuCash, Money Manager Ex, HomeBank, and AbilityCash include the main features as mentioned in Table 1.2.

4.2 The Basic Financial Equation. GnuCash OSS Example

It is presented in the official documentation of GnuCash the basic financial equation (GnuCash, 2022). The equation fundamentally links to all five variables. The fundamental financial equation is:

$$\mathbf{PV*(1 + i)^\wedge n + PMT*(1 + iX)*[(1+i)^\wedge n - 1]/i + FV = 0} \quad (1)$$

(GnuCash, 2022)

Where: X = 0 for end of period payments; X = 1 for beginning of period payments.

Based on this equation, functions can be derived that solve the individual variables. Details are available in the official documentation of the GnuCash source code [41]. Variables A, B, and C are defined for optimized illustration:

$$A = (1 + i)^\wedge n - 1 \quad (2)$$

$$B = (1 + iX)/I \quad (3)$$

$$C = PMT*B \quad (4)$$

$$n = \ln[(C - FV)/(C + PV)]/\ln((1 + i) \quad (5)$$

TABLE 1.2
Main Features of Free & Open Budgeting Software

Free & Open Budgeting Software /License/	Main Features
GnuCash Free licensed under the GNU GPL. www.gnu.org/licenses/gpl-3.0.html Available for GNU/Linux, BSD, Solaris, Mac OS X, and Microsoft Windows.	• Double-entry accounting • Stock/Bond/Mutual Fund accounts • Small-business accounting • Reports, Graphs • QIF/OFX/HBCI Import, Transaction matching • Scheduled transactions • Financial calculations
Money Manager Ex Free foundation software is covered by the GNU Lesser General Public License. www.moneymanagerex.org/about/license Available for Windows macOS, Linux, Android	• Intuitive, simple, fast, clean • Checking, credit card, savings, stock investment, assets accounts • Reminders for recurring bills and deposits • Budgeting and cash flow forecasting • Simple one click reporting with graphs and pie charts • Import data from any CSV format, QIF • Does not require an install: can run from a USB key • Non-Proprietary SQLite Database with AES Encryption • International language support (available in 24 languages)
HomeBank A free personal financial-accounting software, licensed under GNU/GPL Available for Linux, Windows, Mac OS. http://homebank.free.fr/en/index.php	• Cross platform, supports GNU/Linux, Microsoft Windows, Mac OS X • Import easily from Intuit Quicken, Microsoft Money, or other software • Import bank account statements (OFX/QFX, QIF, CSV) • Duplicate transaction detection at import • Multiple currencies, with online update • Automatic cheque numbering • Automatic category/payee assignment • Various account types: bank, cash, asset, credit card, liability • Scheduled transaction, with post in advance option • Transaction template • Category split • Internal transfer • Simple month/annual budget • Dynamic powerful reports with charts • Vehicle cost • Translated in around 56 languages
AbilityCash Free software Available for OS Windows www.softportal.com/software-40921-abilitycash.html	• Work with an unlimited number of accounts • Work with an unlimited number of currencies • Wide choice of transaction details • Hierarchical structure of additional details • Convenient graphic reports • Deferred and repetitive operations, etc.

$$PV = -[FV + A*C]/(A + 1) \tag{6}$$

$$PMT = -[FV + PV*(A + 1)]/[A*B] \tag{7}$$

$$FV = -[PV + A*(PV + C)] \tag{8}$$

The solution for interest is split into two cases.

The simple case for when PMT = 0 gives the solution:

$$i = [FV/PV]^{\wedge}(1/n)-1 \tag{9}$$

The case where PMT!=0 involves an iterative process, which is explained in details in the GnuCash file *libgnucash/app-utils/calculation/fin.c source* (GnuCASH GPL, 2022).

OSS GnuCash gives the following example (GnuCASH, 2022):

How much is the monthly payment on a $100,000 30-year loan at a fixed rate of 4% compounded monthly?

1) The variables must first be defined: n = (30*12) = 360, PV = 100000, PMT = unknown, FV = 0, i = 4% = 4/100 = 0.04, CF = PF = 12, X = 0 (end of payment periods).
2) The second step must convert the nominal interest rate (i) to the effective interest rate (ieff). Since the interest rate is compounded monthly, it uses: ieff = (1 + i/CF)^(CF/PF) – 1, which gives ieff = (1 + 0.04/12)^(12/12) – 1, thus ieff = 1/300 = 0.0033333.

The calculation is: A and B. A = (1 + i)^n – 1 = (1 + 1/300)^360 – 1 = 2.313498. B = (1 + iX)/i = (1 + (1/300)*0)/(1/300) = 300.

With A and B, it is possible to calculate PMT. PMT = –[FV + PV*(A + 1)]/[A*B] = –[0 + 100000*(2.313498 + 1)] / [2.313498 * 300] = –331349.8 / 694.0494 = –477.415296 = –477.42.

Answer: The monthly payments are $477.42.

The mathematically solved calculation of the example is also confirmed by the calculations in OSS GnuCash (Figure 1.4) / N:14–Table 1.3.

Based on the mathematical formula, a number of loan payment calculations were performed, and accurate results were obtained. Figure 1.5 shows visually the interface of the GnuCash Loan Repayment Calculator (access the OSS GnuCash calculator > Tools > Loan Repayment Calculator), where the main parameters of the loan are defined, and the purpose is to calculate the value of the monthly periodic payment and total payments. The configuration is seen in Figure 1.5 (a) Payment Periods: 240 (12 months x 20 years), N:3–Table 1.3, Interest Rate: 3%, Present Value: 110,000, Periodic Payment: Empty and Future Value: 0. Compounding is Monthly, Payments are Monthly, assume End of Period Payments, and Discrete Compounding. Figure 1.5 (b) shows the result: the monthly payment is (a) $ 610.06, and for total period payment (b) $146,414.40.

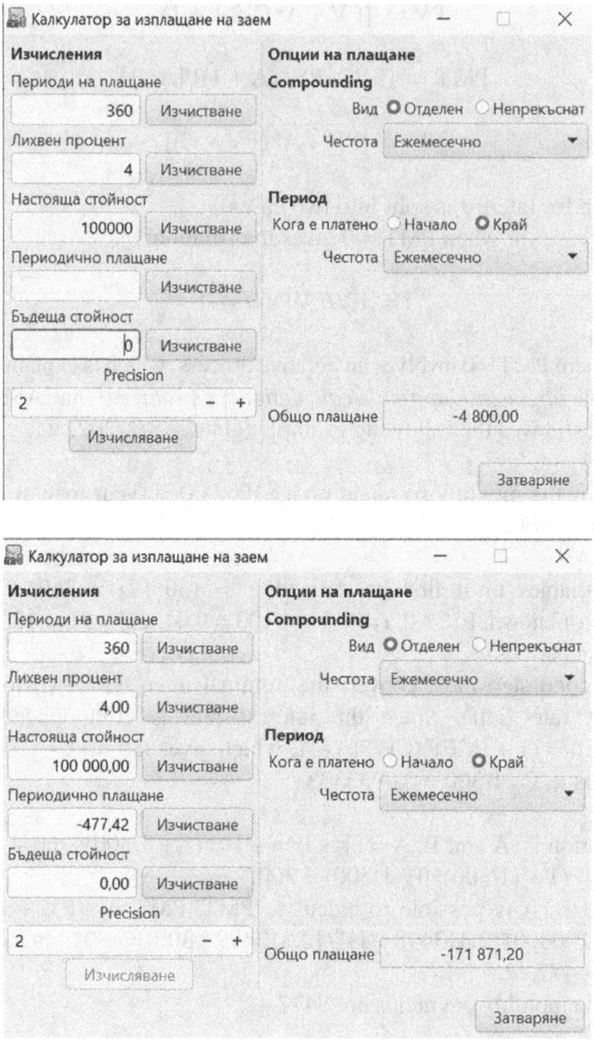

FIGURE 1.4 Accurate result of the loan repayment and confirmed mathematical calculation in OSS GnuCash (a) configuration, (b) result: the monthly payments are $477.42.

Based on the mathematical formula, a number of loan payment calculations were performed, and accurate results were obtained (Table 1.3).

4.3 Financial Data Transfer

While AbilityCash supports financial data transfer in the form of *.xml (Microsoft, 2022), GnuCash, HomeBank, and Money Manager Ex successfully interact via the

TABLE 1.3
Automated Calculations through OSS Gnucash of Monthly and Final Payments on Loans

N:	Payment Periods/ Months 12 *N(Year)	Interest rate %	Present Value $	Future Value	Periodic Payment/ the Monthly Payments $	Total Payment $
1	240	3	100000	0	554.60	133104.00
2	240	4	100000	0	605.98	145435.20
3	240	3	110000	0	610.06	146414.40
4	240	4	110000	0	666.58	159979.20
5	240	3	150000	0	831.90	199656.00
6	240	4	150000	0	908.97	218152.80
7	300	3	100000	0	474.21	142263.00
8	300	4	100000	0	527.84	158352.00
9	300	3	110000	0	521.63	156489.00
10	300	4	110000	0	580.62	174186.00
11	300	3	150000	0	711.32	231396.00
12	300	4	150000	0	791.76	237528.00
13	360	3	100000	0	421.60	151776.00
14	360	4	100000	0	477.42	171871.20
15	360	3	110000	0	463.76	166953.60
16	360	4	110000	0	525.16	189057.60
17	360	3	150000	0	632.41	227667.60
18	360	4	150000	0	716.12	257803.20

QIF file format (FileInfo, 2022). This makes them flexible, interchangeable, and mutually supportive. Striving to ensure the protection of financial data, a good solution is to use imported data or to generate a report in Money Manager Ex (Money Manager Ex, 2021), and in the last stage to store and protect data with encryption (Computerwissen, 2017).

4.4 Generate a General Report/s. Money Manager Ex OSS Example

General reports of Money Manager Ex (MMEX) have MIT license (Money Manager Ex License, 2022).

Permissions include:

- Commercial use;
- Modification;
- Distribution;
- Private use.

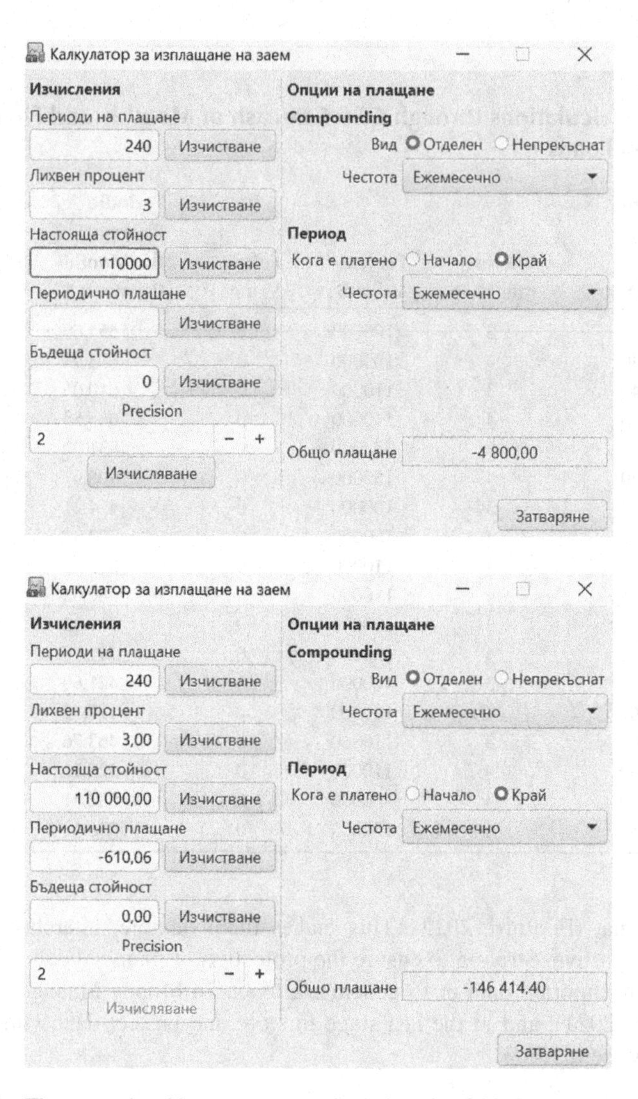

FIGURE 1.5 The example of loan payment calculations by OSS GnuCash for monthly and total payments (a) configuration, (b) result.

Typically, one general report contains (Money Manager Ex Structure, 2022):

1. sqlcontent.sql (MMEX execute the SQL first to return one result set)
   select * from assets
2. luacontent.lua (There are two APIs)
   - handle_record

   function handle_record(record)
   –Your logic to modify a record and apply this function against every record from SQL.

```
record:set("extra_value", record::get("VALUE") * 2);
end
```

- complete

function complete(result)

– Put some accumulated value and apply this function after SQL completes.

```
result:set("TOTAL", 1000);
end
```

3. template.htt (a plain text template) (Money Manager Ex. Html-template, 2022) file powered by
html template which shares the same syntax with Perl's HTML::Template (Peters, 2013)).

4.5 Ensuring the Protection of Financial Data through Encryption MMEX OSS Example

A great convenience and advantage is that MMEX allows the protection of the file directly when encrypted (Figure 1.6). The password is created through (a) "Save" as (b) Encrypted MMB files (Encrypted Database: *.emb).

5 RESULTS

In the process of studying the functional features of budgeting software, technical experiments were performed on generating financial statements, storing data, data transfer, data association, and others. The leading conclusions were formed as follows:

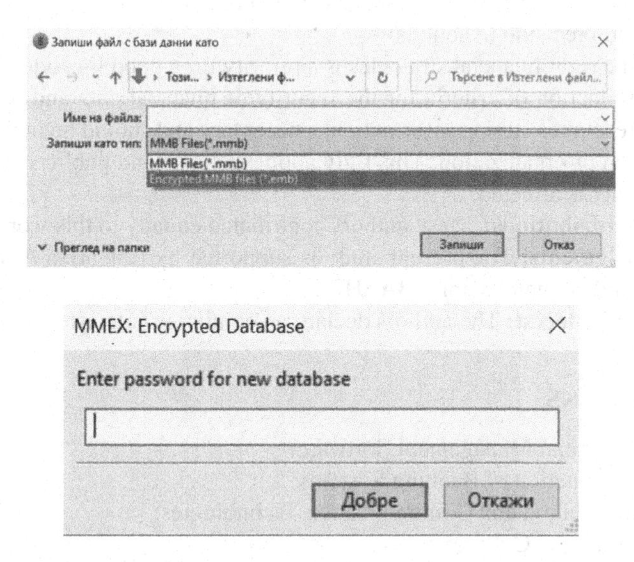

FIGURE 1.6 Creating financial data protection through OSS MMEX (a) encrypting files option, (b) enter password for new database.

- Selected OSS GnuCash provides a built-in financial calculator that optimizes the complex process of bank financial calculations related to loan repayment;
- The selected leading software OSS presents at a high level, and the received financial statements in mathematical values, which are formed in tabular, graphical, and/or visual form (including images, scales, charts, etc.);
- The selected software is largely complementary and given the free access to work can be used simultaneously, it makes the most of their strengths;
- It is possible to store financial data by encrypting directly in OSS MMEX. This optimizes security without the need for additional encryption through external software;
- OSS provide an opportunity to complete the already created concepts or build new ones, using the experience gained.

6 CONCLUSION

This work covers key points needed to build sustainable management and ensure the security of financial data. The implementation of specialized integrated software solutions for financial data management is a correct and successful approach for sustainable financial capital management.

Up-to-date software solutions and tools for personal and corporate banking are presented. The contribution also refers to the public organizations, which are constantly in the process of looking for new opportunities for optimizing the management processes.

The information presented in this report gives, in a compact form, important guidelines to all stakeholders (and this is a large part of the people and society) for choosing and working with modern budgeting software. This leads to the optimization of the management of personal and public finances by integrating constant monitoring and precision in actions.

Building the right financial strategies is a priority for a good lifestyle. At the same time, the introduction of criteria for the security of financial data and the provision of technical resources storing information is necessary and should be in the priorities of every person or organization. This fully applies to state and public systems where national security is affected.

Author Contributions: These authors contributed equally to this work.

Acknowledgments: The present study is supported by Bulgarian Association of Ergonomics and Human Factors (BAEHF).

Conflicts of Interest: The authors declare no conflict of interest.

ABBREVIATIONS

- FMS: Financial Management Software;
- ABS: Automation of Banking Systems;
- ICT: Information and Communication Technologies;
- IC: Information Culture;
- FB: Financial Balance;

- OSS: Open Source Software;
- MMEX: Money Manager Ex.

REFERENCES

Adaptive Planning. Budgeting and planning software for the public sector. (Visited April 2022). www.adaptiveplanning.com/uk/solutions/budgeting-and-planning-software-for-the-public-sector

Aksentijević, N.K.; Ježić, Z. and Adelajda, Z.P. The effects of information and communication technology (ICT) use on human development: A macroeconomic approach. *Economies* 9:128. MDPI 2021. https://doi.org/10.3390/economies9030128

Arellano, A. and Cámara, N. The importance of ICT in society's needs: An empirical approach through Maslow's lens. DIGITAL ECONOMY, BBVA, Financial Inclusion Document. August 2017.

Azhmukhamedov, I.M. *Solving Information Security Problems on the Basis of Systems Analysis and Fuzzy Cognitive Modeling*. Monograph, Astrakhan. 2012.

Barrington, B. *Coordinated Incident Management System (CIMS)*. Third Edition. New Zealand Government. ISBN 978-0-478-43525-2. August 2019.

Cetin Karabat, B.C. and Karabat, C. Increasing awareness of insider information security threats in human resource department. *International Journal of Business and Management Studies*. 4:1. ISSN: 1309-8047. 2012.

Computerwissen. Finanzen im Blick mit Money Manager Ex. 2017. (Visited April 2022). www.computerwissen.de/software/tools/finanz-buchhaltung/finanzen--blick-mit-money-manager-ex/

Council of Ministers of the Republic of Bulgaria (CMRB). Personal Data Protection Act 2018. (Visited December 2021). https://rta.government.bg/images/Image/dpo/zzld.pdf

Cox, A.; Drake, N.; Turner, B.; Wyciślik-Wilson, S. and Clymo, R. Best personal finance software of 2021. *TechRadar*. (Visited April 2021). www.techradar.com/best/best-personal-finance-software.

De Jesus, S.A. Data protection in EU financial services. European Credit Research Institute, ISBN 92-9079-494-1. ECRI Research Report No. 6 April 2004. (Visited December 2021). http://aei.pitt.edu/9429/2/9429.pdf

Egorshin, A.P. *Strategic management: A textbook for universities*. NIMB, Novgorod p. 592. 2009.

European Union Agency for Fundamental Rights and Council of Europe (EUAFRCE). Handbook on European data protection law, 2018. ISBN 978-92-871-9836-5, 2019. (Visited December 2021). www.echr.coe.int/Documents/Handbook_data_protection_BUL.pdf

Feyen, E.; Frost, J.; Gambacorta, L.; Natarajan, H. and Saa, M. Fintech and the digital transformation of financial services: Implications for market structure and public policy. Monetary and Economic Department. BIS Papers, No. 117, July 2021. (Visited December 2021). www.bis.org/publ/bppdf/bispap117.pdf

FileInfo. Quicken interchange format file. (Visited April 2022). https://fileinfo.com/extension/qif

G2. Budgeting and planning software for the public sector: Best public financial management (PFM) systems. (Visited April 2022). www.g2.com/categories/public-financial-management-pfm-system

GnuCash. 8.3. Calculations. (Visited April 2022). https://code.gnucash.org/docs/C/gnucash-guide/loans_calcs1.html

GnuCash. Fin.c source file–description. GNU General Public License (GPL). (Visited April 2022). https://github.com/Gnucash/gnucash/blob/maint/libgnucash/app-utils/calculation/fin.c

Haddon, L. *Information and Communication Technologies in Everyday Life: A Concise Introduction and Research Guide.* Berg, Oxford. 2004. www.researchgate.net/publication/259256921_Information_and_Communication_Technologies_in_Everyday_Life

Hammond, P. and Gummer, B. *National Cyber Security Strategy 2016–2021.* HM Government. (Visited December 2021). www.enisa.europa.eu/topics/national-cyber-security-strategies/ncss-map/national_cyber_security_strategy_2016.pdf

Hashim, A. A handbook on financial management information systems for government: A practitioners guide for setting reform priorities, systems design and implementation. (Based on a compilation of experiences in World Bank-financed projects). *International Bank for Reconstruction and Development / The World Bank.* 2014. (Visited December 2021). www.pfmkin.org/sites/default/files/2020-02/FMIS%20Handbook_Ali%20Hashim_0.pdf

Human Dynamics (HD). Analysis of the national banking legislation from the aspect of personal data protection. Document 1.1.2–8. Final version. (Visited December 2021). https://dzlp.mk/sites/default/files/doc_id_1.1.2.8.pdf

Jang-Jaccard, J. and Nepal, S. A survey of emerging threats in cybersecurity. *Elsevier, Journal of Computer and System Sciences* 80:5:973–993. August 2014. https://doi.org/10.1016/j.jcss.2014.02.005

John, L.Y. and Hoffman, E. Information security policy compliance. 2019. www.researchgate.net/publication/337144310_Information_Security_Policy_Compliance

Karlson, K. Business management trends you should quit in 2019. (Visited December 2021). www.scoro.com/blog/business-management-trends-2017/

Karlson, K. 14 best business budgeting software & tools. Scoro. (Visited April 2022). www.scoro.com/blog/12-best-business-budgeting-software-tools/

Kozhin, V.A.; Shagalova, T.V.; Ivanov, S.A. and Zhestkova I.S. Budgeting. Tutorial. FGBOU VPO Nizhny Novgorod State University of Architecture and Civil Engineering, UDC 338.2, BBK 65.31я73, E40, ISBN 978-5-16-009658-2 (print). 2016.

L.F. 10 best free open source budgeting software for windows. (Visited December 2022). https://listoffreeware.com/free-open-source-budgeting-software-windows/

Microsoft. Open XML formats and file name extensions. (Visited April 2022). https://support.microsoft.com/en-us/office/open-xml-formats-and-file-name-extensions-5200d93c-3449-4380-8e11-31ef14555b18

Money Manager Ex. General reports. (Visited December 2021). www.moneymanagerex.org/features/general-reports

Money Manager Ex. General-reports MIT-License. (Visited April 2022). https://github.com/moneymanagerex/general-reports/blob/master/LICENSE

Money Manager Ex. General-reports report structure. (Visited April 2022). https://github.com/moneymanagerex/general-reports

Money Manager Ex. Html-template. (Visited April 2022). https://github.com/moneymanagerex/html-template

Morgunova, R.V. *Accounting and financial analysis: textbook.* Allowance / R.V. Morgunova, T.V. Kosinets; Vladim. state un-t them. A.G. and N.G. Stoletovs–Vladimir: Publishing house of VlSU, Russia. ISBN 978-5-9984-1123-6., UDC 657 (075.8), BBK 65.053я73. p. 260. 2020.

Nieles, M.; Dempsey, K. and Pillitteri, Y.V. *An introduction to information security.* National Institute of Standards and Technology. June 2017. https://doi.org/10.6028/NIST.SP.800-12r1

Peters, M. Html-template. Metacpan. 21 October 2013. (Visited April 2022). https://metacpan.org/release/WONKO/HTML-Template-2.95/view/lib/HTML/Template.pm

Polyakov, V.P. Aspects of information security in information training. M: FGBNU "IUO RAO". ISBN 978-5-9908256-2-8. p. 135. Moscow. 2016.

Roztocki, N.; Soja, P. and Weistroffer, R.H. *The Role of Information and Communication Technologies in Socioeconomic Development: Towards a Multi-Dimensional Framework. Information Technology for Development,* 2019. ISSN: 0268-1102 (Print) 1554-0170 (Online) Journal homepage: www.tandfonline.com/loi/titd20, https://doi.org/10.1080/02681102.2019.1596654 and www.tandfonline.com/doi/pdf/10.1080/02681102.2019.1596654?needAccess=true

Sahar, A.; Manar A. and Azrilah, A.A. Information security management system. *International Journal of Computer Applications (0975–8887)* 158:7. January 2017. https://doi.org/10.5120/ijca2017912851

Savina, S. and Kuzmina-Merlinob, I. Improving financial management system for multi-business companies. 4th International Conference on Leadership, Technology, Innovation and Business Management. *Elsevier, Procedia–Social and Behavioral Sciences* 210:136–145. 2015.

Schneier, B. *Secrets and lies. Digital security in a networked world.* Wiley Computer Publishing, John Wiley & Sons, Inc. ISBN-13: 978-1119092438, ISBN-10: 1119092434. 2000.

Shakhovskaya, L.S.; Khokhlov, V.V. and Kulakova O.G. *Budgeting: Theory and Practice: Textbook / B98–M.: KNORUS.* p. 400. 2009.

Shchiborsch, K.V. Budgeting the activities of industrial enterprises in Russia. *Publishing House "Delo and Service".* p. 544. 2001.

Tuli, K. and Juneja, N. Cyber security challenges & online frauds on internet. *International Journal of Advanced Research in IT and Engineering.* 5:2. ISSN: 2278-6244. February 2016. www.garph.co.uk

United States Agency for International Development (USAID). *Integrated Financial Management Information Systems. A Practical Guide.* The Louis Berger Group, Inc. and Development Alternatives, Inc. under the Fiscal Reform and Economic Governance Task Order, GEG-I-00-04-00001-00 Task Order No. 06. 2008. (Visited December 2021). www.pempal.org/sites/pempal/files/attachments/PNADK595.pdf

Whitt, P. *Pro Freeware and Open Source Solutions for Business.* APRESS. Winkfield Place, Columbus, GA, USA. ISBN-13 (pbk): 978-1-4842-1131-1, ISBN-13 (electronic): 978-1-4842-1130-4, DOI 10.1007/978-1-4842-1130-4, Library of Congress Control Number: 2015950196. 2015.

Yasenev, V.N.; Dorozhkin, A.V.; Sochkov, A.L. and Yasenev, O.V. *Information Security: Textbook. UDC 311 (075.8).* Nizhny Novgorod, Russia. 2017.

Zakharov, A.N. Corporate security of the company in the context of the development of the Arctic regions of Russia. World civilizations, [online] 1:6. (Visited December 2021). https://wcj.world/PDF/02ECMZ121.pdf (in Russian). 2021.

Zegjda, D.P. and Ivashko A.M. *Fundamentals of Information Systems Security.* Telecom Hotline. 2005.

2 An Efficient Scheme for Detecting and Mitigating Insider Threats

Michael Ekonde Sone and Ange Taffo
University of Buea
Buea, Cameroon

1 INTRODUCTION

To design the security architecture of corporations, governments, and financial institutions, the security administrator must decide on the location and the number of firewalls and intrusion detection systems (IDS) needed. There are internal firewalls and external firewalls associated with the corresponding IDS. While corporations, governments, and financial institutions have designed security mechanisms on external firewalls and IDSs to curb external intrusions, these mechanisms may do little to stop insider threats. These insider threats are because of a lack of effective security mechanisms for the internal firewalls and IDS, which protect the bulk of the enterprise network and most importantly human behavior within the enterprise. There is a need to develop efficient security mechanisms based on human factors for the different components of the IDS, namely the sensor, analyzer, and operator. The sensor and analyzer in the detection mechanism are profile based; that is, they are used to characterize past behavior of users/groups and then detect significant deviations based on analysis of audit records to establish any suspicious event. Meanwhile, the operator, being a human, plays an important role in monitoring, analyzing, and responding to events. The success of the actions of the operator depends on situational awareness, mental workload, and multitasking [1]. This chapter presents an intrusion

DOI: 10.1201/9781003415060-3

tolerant system for detecting and mitigating insider threats. While an IDS aims at identifying threats and ongoing attacks against a monitored system and devises a suitable response, an intrusion tolerant system is equipped with mechanisms for reacting in real time to an ongoing intrusion, striving not just to detect its occurrence, but also to diagnose or predict its evolution so that defensive countermeasures can be invoked timely enough to delay, frustrate, or altogether thwart the attacker's ultimate intentions [2]. Intrusion tolerance within the formalism of Markov Decision Processes (MDPs) is used to determine model parameters namely transition probabilities and cost in each decision stage [2], [3], [4]. The transition probabilities are used to determine the next state from amongst the four possible states in the system, while the cost estimates are based on system operations that are affected while transiting between states. The transition probabilities will be computed using the likelihood principle method based on information-theoretic concepts. The likelihood of insider threats will be determined based on dispositional forces and situational forces. Statistics presented in the ninth annual cost of cybercrime study [5] show that the financial consequences of malicious insiders have increased by 15%. The largest consequence of cybercrime for malicious insiders is business disruption and information loss [5]. Hence, the cost estimates in this research are based on system operations alterations concerning business disruption and information loss. The complete outline of the chapter is as follows. In the next section, the intrusion tolerant system implemented in this research will be presented. In section 3, the causes of insider threats, namely situational forces and dispositional forces, are presented. The basis of the likelihood principle based on information-theoretic concepts is presented in section 4. In addition, the calculation of likelihood is performed for different combinations of situational and dispositional forces obtained from a questionnaire of 100 employees. The cost analysis based on business disruption and loss of information is presented in section 5. Section 6 presents the results used to implement the optimal policy to curb insider threats. Finally, the conclusion and future work are presented in section 7.

2 INTRUSION TOLERANT SYSTEM

An intrusion tolerant system is equipped with mechanisms for reacting in real time to an ongoing intrusion based on the assumption that decisions are made in discrete stages. Each stage begins in a particular state and, probabilistically, transits to a new state (or self-transits to the same state) upon taking control action. The process simultaneously ends the current stage and begins the next. Such a process is a MDP with the outcome of each decision stage defined by the selected control and the realized state transition and associated both with a probability and a cost [2], [3]. Hence, a MDP provides a mathematical framework for modeling decision making in situations where outcomes are partly random and partly under the control of a decision maker (human behavior). The security environment of the MDP implemented in this research is categorized into four states and three control actions as follows [2]:

States

- N: "operating normally"
- A_1: "under attack 1" is considered less severe

- A_2: "under attack 2" is considered severe
- F: "security failure" is a complete failure of the system

Control actions

- W: "wait" action
- D: "defend" action
- R: "reset" action

The four-state Markov model for the different control actions with the associated probabilities and costs is shown in Figure 2.1 [2].

The different cost is estimated in terms of information loss and business disruption when transiting from one state to another for the different control actions. Meanwhile, the probability p_A relates to the frequency with which intrusion attempts are initiated, where larger values pertain to more likely attackers. Probability p_F relates to the rate at which an initiated intrusion will penetrate the security system, where larger values pertain to more vulnerable systems. Probability p_D relates to the success rate of a well-timed defensive countermeasure in preventing an otherwise inevitable security failure, where larger values pertain to more effective response devices [2], [3].

The proposed model will be categorized using three types of responses:

- Instant Response: The mind of the user is gradually changing to hack the company based on some human behavior discussed later in the chapter. At this stage attack 1, A_1, occurs but the attack is back to normal within the shortest possible time hence leading to an instant response by the operator who monitors and analyzes the system. The probability for the attack P_{A1} to occur is high and the cost is low hence the operator is alert.

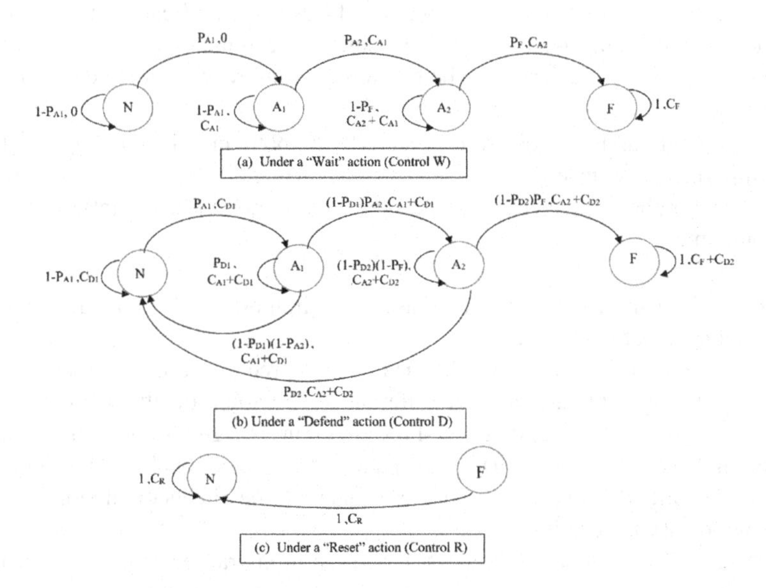

FIGURE 2.1 Four-state and three-action Markov model.

- Reversible Response: The mindset of the user has changed completely to an intentional and malicious user hence a significant change in human behavior which will cost the company more time to mitigate the attack due to the attack complexity. The probability for the attack P_{A2} to occur is lower and the cost is higher than for the instant response.
- Irreversible Response: At this stage, the attack causes permanent damage, and the organization needs to shut down and reset operations. The cost to reset is very high and the probability of the attack is very low, which made it difficult for the operator to mobilize the necessary resources to curb the attack.

3 CAUSES OF INSIDER THREATS

Insider threats could result due to two main forces, namely [6]:

- Situational forces
- Dispositional forces

Situational forces are based on provocation or threat and reward of an employee. Meanwhile, dispositional forces are based on human behavior. Hence, threats that appear to have been prompted by provocation or other situational forces should be perceived as caused by the situation rather than by the person. The two forces are subjected to the logic of the discontinuity principle that operates hydraulically [6]. The logic is such that, as the strength of the situational causality increases, dispositional causality is assumed to grow weaker and vice versa. Based on motives, situational forces could be classified as situational provocation or reactivity and situational rewards or instrumentality [6].

Dispositional forces based on dispositional inference find causes that underline human behavior. Human behavior traits considered are attitudes, ability, and morality. Morality is considered separately, whereby morality inferences could be stated as [6]:

- Expect moral behavior from persons with both moral and immoral traits, implying more uncertainty.
- Expect immoral behavior from persons with immoral traits, implying a certain outcome.

Hence, dispositional inferences based on immoral behavior are relatively unaffected by situational forces.

In this research, we have established a new tree diagram used to establish inferences on insider threats based on different combinations of situational forces and dispositional forces. The tree diagram shown in Figure 2.2 grows downwards with the branches indicating the causes. The tree diagram is conceived such that insider threats are fundamentally due to human behavior traits with the effects of situational forces being an increase in severity.

Note that dispositional inferences based on immoral behavior are relatively unaffected by situational forces, hence no branches are emanating from immoral traits.

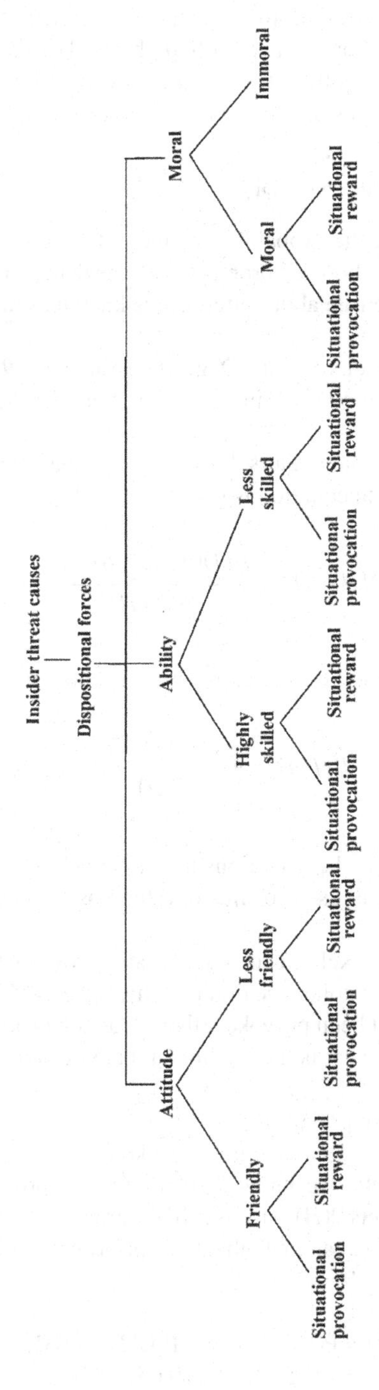

FIGURE 2.2 Insider threats causes tree diagram.

4 LIKELIHOOD PRINCIPLE

In this section, we shall present the basics of the likelihood principle, which establishes inferences based on an a priori hypothesis. The likelihood principle will be used to establish inferences from the hypothesis established from a questionnaire administered to 100 employees of some selected companies.

4.1 BASICS OF LIKELIHOOD PRINCIPLE

The likelihood is based on the probability of the outcome that happened given the different hypotheses; that is, how the probability of the data that happened varies with the hypothesis. This simple rule about inference is known as the likelihood principle and is stated as follows [7]:

Given a generative model for data D given parameters $\boldsymbol{\theta}$, $P(D|\boldsymbol{\theta})$, and having observed a particular outcome D_1, all inferences and predictions should depend only on the function $P(D_1|\boldsymbol{\theta})$.

If θ denotes the unknown parameters, D denotes the data, and H denotes the overall hypothesis space, the general equation is given as:

$$P(\theta|D,H)=\frac{P(D|\theta,H)P(\theta|H)}{P(D|H)} \tag{1}$$

Using Bayes' theorem and known parameters, (1) could be written as follows:

$$P(H|D)=\frac{P(D|H)P(H)}{P(D)} \tag{2}$$

(2) is the basis of likelihood calculations in this research.

The following example is used to illustrate the basic calculation of likelihoods based on the hypothesis.

Suppose that a system is likely to come under attack by $1/10^{th}$ of all employees. If you have been provoked by a colleague or hierarchy, there is 95% of an attack and 5% for no attack. If you haven't been provoked, then an attack is possible 5% of the time. Suppose there is an attack, what is the likelihood that you were provoked?

D = data: attack

H = hypothesis: you were provoked

H' = the other hypothesis: you were not provoked

Before acquiring the data, we know that the a priori probability of causing an insider attack is 0.1 which sets P(H). This is called a priori. We also need to know P(D).

From the sum rule, we can calculate that the a priori probability P(D) of an attack, whatever the truth may actually be is:

$$\begin{aligned} P(D) &= P(D|H)\,P(H) + P(D|H')\,P(H') \\ &= (0.95)\,(0.1) + (0.05)\,(0.9) \\ &= 0.14 \end{aligned}$$

And from (2), we can conclude that an employee performing an insider attack given he was provoked is given as:

$$P(H|D) = \frac{P(D|H)P(H)}{P(D)} = \frac{(0.95)(0.1)}{0.14} = 0.68$$

4.2 Calculation of Likelihood

The hypotheses were established for the different branches in Figure 2.2 using a questionnaire filled by 100 employees from some selected small and medium-sized companies in Buea and Douala in Cameroon, West Africa.

Data on the dispositional forces of the employees was obtained from the company administration, while that on situational forces was obtained directly from the employees. The results are shown in Figure 2.3.

In Figure 2.3, the administration gives the number of employees with the human behavioral traits such as 60 employees is friendly while 40 are less friendly. For the insider threats data, employees in the different categories of human behavioral traits are asked about their susceptibility to perform insider attacks. For example, 20% of the friendly employees could perform an insider attack. In addition to the dispositional forces, situational forces could increase the inference of the threat, for example, 30% of the 20% of the friendly employees could perform insider attacks when provoked. It is worth noting that our survey discovered that 80% of the friendly employees would not perform insider attacks even under situational forces. The same hypothesis is used for all the other human behavior traits.

(2) is used to calculate the likelihood for the different scenarios of the hypotheses in Figure 2.2. The results of the different probabilities are displayed in Table 2.1.

From Table 2.1, it is seen that an employee who is highly skilled and expects a reward is most likely to perform an insider attack.

5 COST ANALYSIS

As mentioned in [5], the largest consequence of cybercrime for malicious insiders is business disruption and information loss. Hence, the cost estimates in this research are based on system operations alterations concerning business disruption and information loss. The different classes of the cost will be explained and grouped into business disruption or information loss.

Cost can be classified as direct/tangible cost and indirect/intangible cost. Direct/tangible costs involve financial losses and asset losses. These represent the monetary value of all services, hardware, software, and other resources in cybersecurity systems. Meanwhile, indirect/intangible costs depend on investment decisions such as cybersecurity breaches. For example, if an organization has a widely known breach, it could lose current or future customers because of the effects on its reputation. Hence, indirect/intangible cost is mainly business disruption. Some different types of costs concerning business disruption and information loss for direct/tangible costs are as follows:

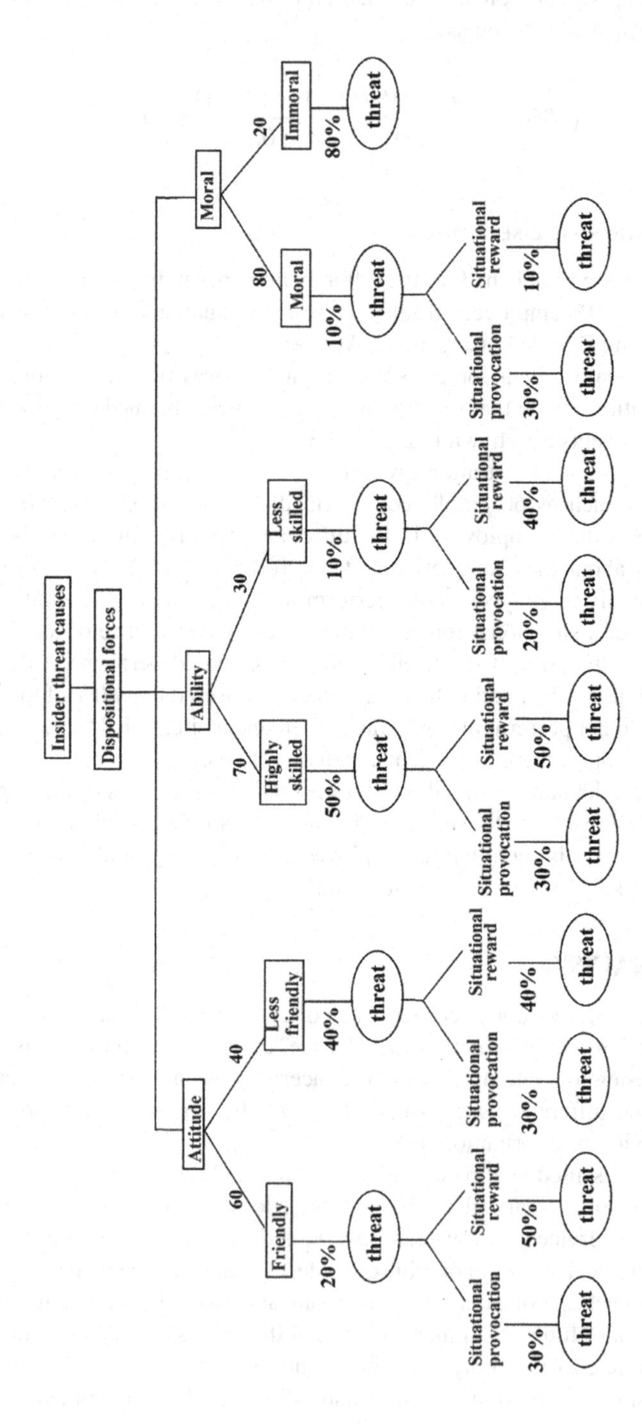

FIGURE 2.3 Hypotheses tree for insider threats.

TABLE 2.1
Probabilities for the Different Dispositional/Situational Forces Scenarios

S/N	Different Scenarios	P(H)	P(D | H)	P(D | H′)	P(H′)	P(D)	P(H | D)
1	FRIENDLY	0.6	0.2	0.8	0.4	0.44	0.27
2	FRIENDLY & PROVOKED	0.6	0.5	0.5	0.4	0.5	0.60
3	FRIENDLY & REWARD	0.6	0.7	0.3	0.4	0.54	0.78
4	LESS FRIENDLY	0.4	0.4	0.6	0.6	0.52	0.31
5	LESS FRIENDLY & PROVOKED	0.4	0.7	0.3	0.6	0.46	0.61
6	LESS FRIENDLY & REWARD	0.4	0.8	0.2	0.6	0.44	0.73
7	HIGHLY SKILLED	0.7	0.5	0.5	0.3	0.5	0.70
8	HIGHLY SKILLED & PROVOKED	0.7	0.8	0.2	0.3	0.62	0.90
9	HIGHLY SKILLED & REWARD	0.7	1	0	0.3	0.7	1.00
10	LESS SKILLED	0.3	0.1	0.9	0.7	0.66	0.05
11	LESS SKILLED & PROVOKED	0.3	0.3	0.7	0.7	0.58	0.16
12	LESS SKILLED & REWARD	0.3	0.5	0.5	0.7	0.5	0.30
13	MORAL	0.8	0.1	0.9	0.2	0.26	0.31
14	MORAL & PROVOKED	0.8	0.4	0.6	0.2	0.44	0.73
15	MORAL & REWARD	0.8	0.2	0.8	0.2	0.32	0.50
16	IMMORAL	0.2	0.8	0.2	0.8	0.32	0.50

- Cost of repair
- Lost productivity
- Value of damaged equipment
- Value of lost data
- Cost to replace equipment
- Cost to reload data

Cost is estimated based on the impact on the organization. Based on the National Institute of Standards and Technology (NIST) specifications, the impact could be classified as shown in Table 2.2 [8].

The cost for the different stages and the different control actions will be calculated using a risk level matrix and risk scale as displayed in Table 2.3 [8].

6 RESULTS TO IMPLEMENT OPTIMAL POLICY

The cost for the different transitions and the different control actions will be calculated. The costs will be associated with the transition probabilities to propose optimal policy for the different responses. The cost for the different control actions is

TABLE 2.2
Magnitude of Impact and Corresponding Definition

Magnitude of Impact	Impact Definition
High	(1) may result in the high, costly loss of major tangible assets or resources. (2) may significantly violate, harm, or impede an organization's mission, reputation, or interest; or (3) may result in human death or serious injury.
Medium	(1) may result in the costly loss of tangible assets or resources. (2) may violate, harm, or impede an organization's mission, reputation, or interest; or (3) may result in human injury.
Low	(1) may result in the loss of some tangible assets or resources; or (2) may noticeably affect an organization's mission, reputation, or interest.

TABLE 2.3
Risk Level Matrix and Risk Scale

	Impact		
Threat Likelihood	Low (10); Medium (50); High (100)	Low (10); Medium (50); High (100)	Low (10); Medium (50); High (100)
High (1.0)	Low 10 x 1 = 10	Medium 50 x 1 = 50	High 100 x 1 = 100
Medium (0.5)	Low 10 x 0.5 = 5	Medium 50 x 0.5 = 25	High 100 x 0.5 = 50
Low (0.1)	Low 10 x 0.1 = 1	Medium 50 x 0.1 = 5	High 10 x 0.1 = 10

calculated with the hypothesis such that under the "wait" action, the row of the risk level matrix is "low"; that of the "defend" action is "medium". In addition, any transition to the system failure F state is "high". The complete calculation of the cost for the different control actions and the four states are shown in Table 2.4.

The optimal policy in our model is such that the system begins under the "wait" action. If there is an attack, the system reverts to the "defend" action. If the attack is complex and not addressed within the stipulated time it reverts to the "reset" action. The goal is to determine threshold values for cost and transition probabilities, which will enable the system operator in the IDS to prevent the system from transiting to F under the "wait" action and to revert to the "defend" action, and under the "defend" action conditions should be such that system transiting to F should be minimized while A_1 or A_2 transiting back to N should be maximized. The results in Table 2.1

TABLE 2.4
Cost for Different Control Actions

Control W			Control D					Control R
C_{A1}	C_{A2}	$C_{A1}+C_{A2}$	C_{D1}	$C_{A1}+C_{D1}$	$C_{A1}+C_{D2}$	$C_{A2}+C_{D2}$	$C_{F}+C_{D2}$	C_{R}
1	10	5	10	25	50	100	100	100

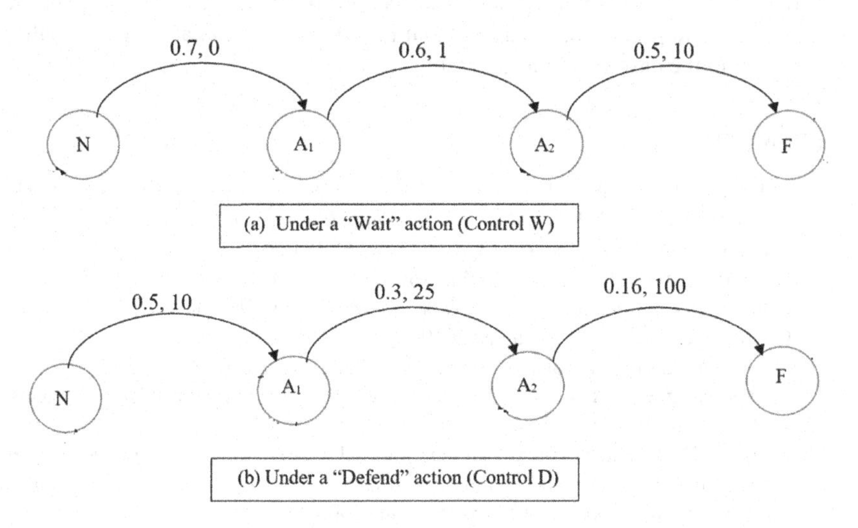

FIGURE 2.4 Threshold values for an operator to mitigate insider threats.

and Table 2.4 are used to state the threshold values for the different transitions that the operator will use to mitigate insider threats. The rationale for establishing the threshold values is such that the highest probability and lowest cost will correspond to a high index in the anomaly-based approach in the IDS and will provoke an instant response from the operator. For example, N to A_1 transition with a transition probability of $P_{A1} = 1.00$ which represents a highly skilled/reward employee will trigger a "defend" action. For effective mitigation, the probability $(1\text{-}P_{D1})$ $(1\text{-}P_{A2})$ should be high compared to $(1\text{-}P_{D1})P_{A2}$. In our implementation, the forward transition probabilities will be proposed as thresholds for the operator to determine the response to mitigate the insider attack. Meanwhile, the reverse probabilities are based on the system capabilities at the disposal of the operator to revert to normal operations or avoid system failure. Figure 2.4 shows threshold values for the "wait" action and the "defend" action.

It should be noted that, as the probability increases the cost reduces, and under the "defend" action the probability values are small implying that dispositional or situational forces that could lead to insider attacks are not perceived by an operator as such.

7 CONCLUSION AND FUTURE WORK

In this chapter, we have presented an efficient system that could detect and mitigate insider threats based on human behavioral traits and situational forces. Threshold values were proposed to enable the operator in an IDS to determine an insider threat and perform the appropriate response. It is shown that for a threat that involves a highly skilled employee expecting a reward for the attack, the response is instantaneous. Threshold values are forward transitions that are based on human behavioral traits. In future work, reverse transition probabilities based on system capabilities will be studied and proposed. In addition, an FPGA-based automated system will be implemented in the case of an instant response.

REFERENCES

[1] P. Lif and T. Sommestad, "Human factors related to the performance of intrusion detection operators", *Proceedings of the Ninth International Symposium on Human Aspects of Information Security & Assurance (HAISA 2015).*

[2] O. Patrick Kreidl, "Analysis of a Markov Decision Process Model for intrusion tolerance", *Dependable Systems and Networks Workshops (DSN-W)*, 2010 International Conference, IEEE Xplore, August 2010.

[3] B. Madan, K. Goševa-Popstojanova, K. Vaidyanathan and K. S. Trivedi, "A method for modeling and quantifying the security attributes of intrusion tolerant systems", *Performance Evaluation*, 56:167–186, 2004.

[4] K. Joshi, W.H Sanders, M.A. Hiltunen and R.D. Schlichting, "Automated recovery using bounded partially observable Markov decision processes", *Proc. of Dependable Systems and Networks (DSN)*, 445–456, June 2006.

[5] K. Bissell and L. Ponemon, "Ninth annual cost of cybercrime study", *Accenture*, 2019.

[6] G.D. Reeder, S. Kuramn, H.S. Hesson-McInnis and D. Trafimow, "Inferences about the morality of an aggressor: The role of perceived motive", *Journal of Personality and Social Psychology*, 83:4:789–803, 2002.

[7] D.J.C. Mackay, *Information theory, Inference, and Learning Algorithms*, Cambridge University Press, 2003.

[8] J. Davis, "Formula for calculating cyber risk", *State of Security, General Infosec*, 2021.

3 Phishing Through URLs: An Instance Based Learning Model Approach to Detecting Phishing

Gerardo I. Armenta and Palvi Aggarwal
Department of Computer Science, The University of Texas–El Paso, USA
El Paso, Texas, USA

1 INTRODUCTION

With the Internet now being a necessity for people to communicate, it has turned into an indispensable commodity. With great ideas and technological advances, abuse and malicious intent come as well, as some people intend to use such technologies for their benefit. Phishing was first recorded to occur on January 2, 1996, though phishing attacks occurred years before when America Online (AOL) was the top Internet service provider (ISP) in the United States (KnowBe4., 2017). While the start of illicit methods of gaining private information started with getting access to users' accounts through guessing credit cards or the use of warez, phishing methods have changed since then. Moving between targeting people by email to attackers passing off as corporate employees or friends, all the way to gaining social media user interest and guiding their victims to visit a spoofed website address to steal their information.

DOI: 10.1201/9781003415060-4

Phishing in social media has been increasing throughout the years since the majority of people have at minimum one social media account. Out of the 5 billion Internet users worldwide, approximately 4.65 billion are active social media users (Statista, 2022). This has enabled attackers who would target people with phishing emails to also exploit social media sites and target even more people. Due to this, social media sites have taken steps to mitigate these types of attacks but unfortunately it is not perfect, and some attacks do get to bypass their defenses. Due to the animosity of attackers and the small percentage of risk there is of getting caught, phishing is for the most part a risk only to the attacker's target which turns into the victim.

Extensive research has been done to fit the best methods to train the end-user to identify phishing emails, but not much has been done in the realm of social media (Cranford et al., 2019). Just to put things into perspective, $2.3 billion in damages has been caused by phishing attacks from 2013 to 2016 (Ubing et al., 2019). There are several attempts being made to deter phishing, but unfortunately some phishing is still occurring. In this research, we will focus on what people can do to prevent falling victims to such attacks without the use of technology. This aims to reduce the likelihood of being preyed on and the likelihood of falling victim without the aid of current technologies.

The goal of the Instance Based Learning (IBL) model is to act and copy the behavior of a human when presented with the decision of whether to click on a URL link. This can then provide insights on whether lexical training can help people in reducing the likelihood of falling victim. Moreover, there will not be a focus on the current techniques used to prevent phishing in social media sites since there is plenty of research in that realm and this research wants to focus on the human aspect of what can be done.

2 RELATED WORK

There is an increase in research being done to counter phishing attacks. To that extent, there is some research being done to better understand the behavioral aspect of human decision making when presented with phishing attacks. The main focus on using IBL models in phishing includes the cognitive modeling that intends to represent the human behavior from start to end on decision-making processes in the classification of emails.

Shonman et al. (2018) identified cues in emails to identify as legitimate or illegitimate. Their model represents how boundedly rational humans make decisions; the more cues that are present, the more likely an email is considered phishing. The results show the importance in text similarity for accuracy and by using an IBL model. The predictions are made to see how humans will respond to emails based on the experiment. These experiments use language processing methods such as LSA, GloVe, and Bidirectional Encoder Representation from Transformers (BERT) (Briggs, 2021) to embed the words found in the emails. The embeddings were used in this experiment to calculate and represent the similarity between two email instances for the IBL model. An average 79.7% accuracy determined that the model was able to predict how the user would react to that email (Shonman et al., 2018).

James et al. (2013) utilized several machine learning classifiers to categorize URLs as phishing demonstrating the success rate of each. Efficiency in utilizing machine learning is high in detecting phishing URLs, which leaves about less than 10% not being detected.

Alshira'H et al. (2020) provide different machine learning and statistical models used to classify URLs. These technologies are used by social media sites as a measure to identify malicious URLs being used for phishing. The decision tree classification algorithm is a machine learning algorithm that applies the divide and conquer approach by using recursive partitioning. Another machine learning algorithm is the random forest classifier that uses ensemble learning methods and consists of multiple decision trees. This is done to reduce bias and variance on the decisions.

The reason to use IBL modeling instead of machine learning algorithms is because many social networks are currently implementing such machine learning algorithms but still have a margin of error where some phishing URLs aren't detected. Thus, the IBL model intends to provide an insight into whether a person can detect malicious URLs and to what degree.

A cognitive model that is based on IBL can be used to classify malicious URLs. This will allow the verification of important cues to be identified in a URL and how with noncomplex methods, a person can reduce their exposure to phishing threats. To determine if we can improve our detection of phishing URLs without the help of technology, we determine a reasonable set of parameters to focus on in a URL to make that determination.

Next, we ask ourselves what the percentage of improvement is when identifying such flags in a URL to make a reasonable assessment. We must keep in mind that not everyone is tech savvy in the domain of URLs and that trying to teach this material to help people make informed decisions will make uninterested parties disregard it.

1. How do people use similarity to classify URLs?
2. What kind of training do people need to get for improving malicious URL detection?
3. Simple vs complex similarity??

3 TASK DESCRIPTION

In this paper, we present how a cognitive model that is based on IBL can be helpful to identify malicious URL that people can also do. This will also help identify important queues to look for in a URL, and how with noncomplex methods, a person can reduce their exposure to phishing threats.

3.1 URL Component

We will focus on the hostname of a URL and divide it into three parts as shown in Fig. 3.1.

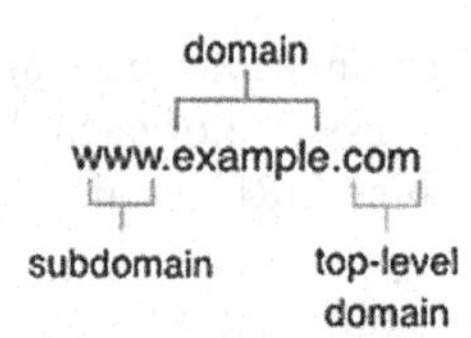

FIG. 3.1 The hostname of a URL is Safe URL composed of three elements: the subdomain, domain, and the top-level domain. (Ubing et al., 2019).

The hostname is the element that will be used for the IBL model to make its classifications. A URL consists of multiple parts, but the hostname is a feature that remains consistent throughout. The majority of websites utilize the hostname to get to their home web page. While web pages utilize a protocol, namely the Hypertext Transfer Protocol (HTTP) or Hypertext Transfer Protocol Secure (HTTPS), in the majority of web pages this doesn't change unless there is something very specific, which would use a different protocol and would be used for a very specific purpose. As such, targets will fall victims to an attack so we will not make considerations for this parameter. It is important to note that the majority of subdomains use HTTP or HTTPS.

The port may follow the hostname, although it isn't common. Since the port is not needed in a URL to work, we exclude this part from the model. Lastly, the path follows a URL. The path is typically used for the different pages available from the hostname to show different information. Nevertheless, a path is optional and tends to be a page from the domain of the hostname. As such, this part is also not used in the model.

3.2 DATA

Extraction of URLs come from 100 of the most popular websites accessed in the United States (Semrush, 2022). This list contains URLs that are not only visited the most but are also shared through social media. One hundred known and verified malicious URLs were also extracted for the model (PhishTank, 2022). Additionally, there will be 100 URLs that are very similar to those that are safe to mimic phishing sites that try to prey on error from the user due to the likeliness of safe and popular sites. This data will be used by the IBL model to test our hypothesis. See Fig. 3.2.

It is important to use current data to pick up the current trends being used by malicious actors. Phishing has evolved and different tactics are used to make targets fall victim to an attack. It is important to note that subdomains are starting not to be needed to access a web page. Many phishing URLs are also taking advantage of this feature. For example, entering google.com on chrome will automatically take you to www.google.com and it is the same for phishing URLs. While this feature is being used by many domains, the subdomain in phishing sites is in some cases different than the typical www. As such, this part of the hostname is important to determine if a URL is malicious or not.

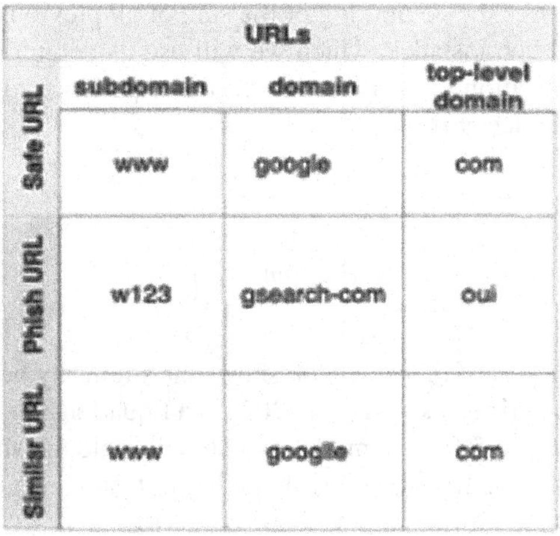

URLs			
	subdomain	domain	top-level domain
Safe URL	www	google	com
Phish URL	w123	gsearch-com	oui
Similar URL	www	googlle	com

FIG. 3.2 URLs for classification.

Some of these sites extracted did not feature a subdomain when obtained. The World Wide Web subdomain (www) was added to those that were missing this section in the URL. This added complexity to the model's decision making since it compared these three parts of an URL before making a decision.

4 INSTANCE BASED LEARNING MODEL

We developed an IBL cognitive model that makes a choice of being presented with a URL (Lejarraga et al., 2012). This model makes decisions based on memory mechanisms from the ACT-R cognitive architecture (Xu et al., 2022). The decisions made by the model are then stored in memory making each past decision an instance in memory. The activation for an instance in memory is as follows in Equation 3.1.

Equation 3.1

$$A_{i,k,s} = ln \sum_{t_i = 1,t-1} (t - t_i)^{-d} + MP \sum Sim(v_k, c_k) + \sigma^* ln\left((1 - \gamma_{I,k,t}) / \gamma_{I,k,t}\right)$$

The activation A takes the base activation adding the partial matching of the similarity between the current instance and the ones in memory. Equation 3.2 shows the base activation part of the activation equation and gets the previous instances with the decay in memory throughout time.

Equation 3.2

$$ln \sum_{t_i = 1,t-1} (t - t_i)^{-d}$$

The base activation uses the natural log of the sum of all previous instances, t_i to t–1 and d, the decay of such instance, which we will use its default value of 0.5 in this model. The base activation takes into account the experience of forgetting as time passes, which is what decay is used for.

Equation 3.3

$$MP \sum_{k} Sim(v_k, c_k)$$

The partial matching part of the activation shows the similarity between the previous instance, vk, and the current, ck (Xu et al., 2022), in Equation 3.3. The mismatch penalty is by default set to 2.5 for our model, which will scale the similarity value. The similarity value will range from 0 to 1 with the highest value meaning that the values are equal and 0 being that the values have no similarity.

Equation 3.4 represents the randomness of decision making. It allows for the model to not stay with a specific choice and to explore different choices. The equation includes 1–gamma over gamma, where gamma takes a random number from 0 to 1 and the result then takes the natural log of the value times the variance that is the noise in the function. The noise is set to a default of 0.50 for this model.

Equation 3.4

$$\sigma * ln\ ((1 - \gamma_{I,k,t}) / \gamma_{I,k,t})$$

The IBL model will make a choice based on the highest blended value it gets. This value is a computation that will consider the probability of retrieval multiplied by the outcome for an option k at trial t (Xu et al., 2022). The equation is represented in Equation 3.5.

Equation 3.5

$$V_j = \sum_{i=1}^{n} P_{ji} * X_{ji}$$

The blended value is computed once all the activations are completed (Carnegie Mellon University, 2022). The variable P is the probability of retrieval of instance i for option j. The outcome of an instance is represented by X of option j of instance i.

4.1 Simulation Task Using IBL Model

We utilize PyIBL for the development of the IBL model. The choices of the URLs are randomized for each trial and the same URLs might be presented more than once per trial due to the randomness of the model. This helps to assess if the model is learning from its past experiences.

The URL that is presented to the model to classify is transformed into the three host-named components: subdomain, domain, and top-level domain. Each of the components are presented to the similarity function of the model independently but decides by taking all the components into consideration. There are two similarity functions that are used by the model, BERT, with cosine similarity and lexical based similarity named Basic Comprehensive Similarity.

Situation: Define the attributes

2 x 4 design

2–similarity types
4–pre-population conditions

The model is trained in four different ways for four different experiments with 30 safe URLs and 15 phishing URLs, 30 safe URLs and 15 similar to the safe URLs, 10 safe URLs and 0 phishing URLs, and 10 safe URLs with 0 similar to the safe URLs. These URLs and the labels of either being safe or not are pre-populated to the model before the initial trial begins.

The four different pre-population methods serve to represent the following two scenarios: 30 safe and 15 phishing or 30 safe and 15 similar, is to be considered as having conceptual knowledge of some URLs and basic knowledge of malicious URLs; as for the 10 safe and 0 phishing or 10 safe and 0 similar, is to be considered as the limited knowledge of URLs with understanding of some popular websites but has no concept of malicious URLs.

Model participants = 100
Trials for each participant = 200

When the 200 trials start, there are 200 URLs that the randomizer can choose from. Once the randomizer picks a URL, the choice is then presented to the agent to determine if it is safe or malicious. It is important to note that there are 100 safe URLs, 100 phishing URLs, 100 similar to the safe URLs.

Define action
Define utility for each action

The hostname is split up into three categories, the subdomain, domain, and top-level domain. Each category is compared with each other in the same category for similarity so the agent can decide on the URL that is presented. It takes the hostname components individually to make an informed decision based on the pre-populated data and its experience gained after each trial.

4.2 Similarity Functions BERT

BERT is a Natural Language Processing model (NLP) that was released by Google in 2017 (Rogers et al., 2020). It consists of transformers that are made up of encodings

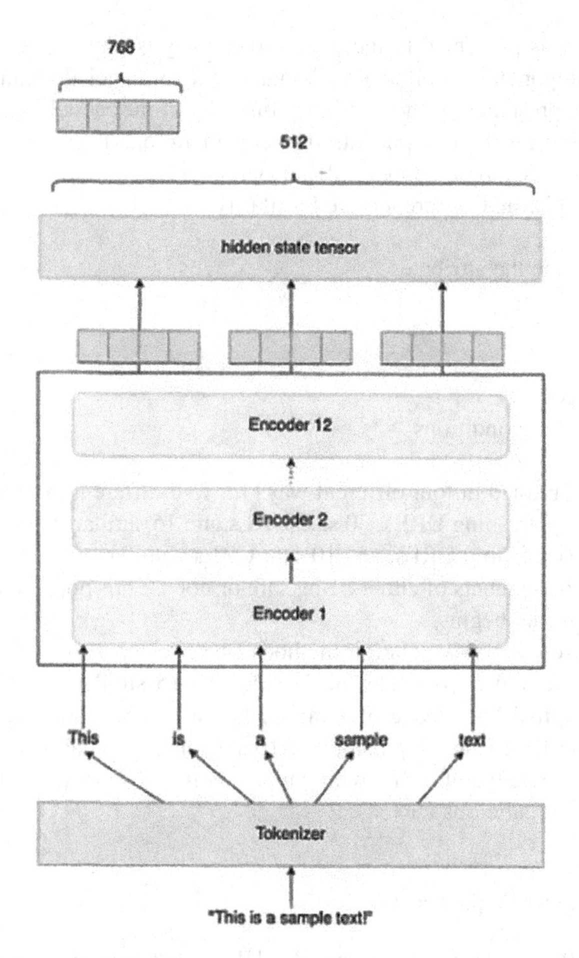

FIG. 3.3 "This is a sample text!".

that generate 512 tokens with a length of 768 per each word. As shown in Fig. 3.3, BERT uses encoders on tokenized words to create 512 tokens of 768 in length that are called tensors. In our model, we will take the mean of the tokens to input only 128, half of the original. Since we will be dealing with only one word per text similarity, taking the mean to reduce its size is negligible. Everything else will remain the same when using BERT to get the similarity.

BERT uses semantic similarity, which means that it will take a word and compare it to others based on synonym and hyponyms. This means that it will take higher level encoding for the similarity of each part of the hostname **rather than** the lexical similarity found in our custom use. Cosine similarity will be used on the embeddings to obtain the similarity value in the range from 0 to 1, with 1 being exactly the same in similarity.

4.3 Basic Comprehensive Similarity

We introduce a similarity function which helps associate the likeliness of the current hostname component and the previous ones in memory. It compares the two components by taking character by character and assigns a "1" if alike or a "0" if not. The total is then divided by the largest URL component's length to get the similarity between both components to get a similarity value between 0 and 1. Fig. 3.4 shows how the function worked to calculate the similarity value between two components.

The 0.67 shown in Fig. 3.4 is the similarity value assigned to the current component with the past one. This is done for all three components of the hostname. All three components of the hostname are weighted equally, and the agent can make an informed decision based on that information.

5 SIMULATION PROCEDURE

One actor was used to partake in 100 scenarios equating to 100 participants. Then, each of the participants completed 200 trials. We pre-populated the model in four different runs to obtain a better understanding of the results. We made one run with a pre-population of 30 safe URLs and 15 phishing and another with 10 safe and 0 phishing. This allowed different levels of knowledge a user has to be represented when dealing with URLs. The other two runs are composed of the following: one is with pre-population of 30 safe URLs and 15 similar URLs, while the other one is 10 safe and 0 phishing. This allowed different levels of knowledge a user has to be represented when dealing with URLs. The other two runs are composed of the following: one is with pre-population of 30 safe URLs and 15 similar URLs, while the other one is 10 safe and 0 similar.

All the models were pre-populated with a reward of 10 for asserting that a phishing URL was not safe and for selecting a safe URL was indeed safe. In contrast, "-10" points were given as a punishment in the pre-population for the latter, which would reflect that the participant chose wrong. In the 200 trials that were given to each participant, 100 points in reward was given for a correct response for choosing safe on a safe URL and not safe for a phishing or similar URL. The participant was penalized with "-100" for selecting a safe URL and not safe for a phishing or similar URL. The

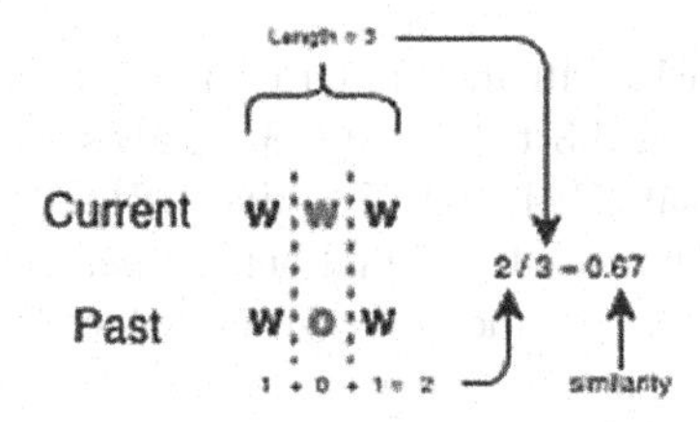

FIG. 3.4 Similarity value assigned to the current component.

participant with "-100" for selecting a safe URL that had phishing, while they were penalized with "-500" for selecting safe on a phish URL. This was done to best interpret the potential loss when clicking on a phishing URL, while there is no loss for not selecting a safe URL. From there, the information was captured to identify the accuracy of the model and if there was improvement in detecting phishing correctly. We had to check and test not only on the phishing detection but if the model was marking safe sites as phishing.

The similarity function was tested to confirm no similarity value assigned was over 1 or less than 0 as that would automatically convert the similarity to either 0 if it was less than or 1 if it was greater than. Even though the IBL model would convert the similarity to these values, we wanted the similarity value computed to fall within the continuous range of 0 to 1.

6 RESULTS

The results provided by the IBL model provided an insight into the learning capabilities possible to make an informed decision on a URL. The model created a bias in that when two of the components in the hostname had a high similarity value it would select as safe regardless of what the other component was. This varied on the subdomain and with the top-level domain. This was due to the phishing URLs having a subdomain of www and/or the top-level domain being com. Whenever one or the other was present, it would tend to believe it is a safe URL. The model did learn from its experience and would not select the site as safe after learning it was not in a previous instance.

6.1 Classification Accuracy

When running BERT similarity on the model, the results show a 38% accuracy on selecting appropriately when a phishing URL is presented. Selecting safe when a safe URL is present is right below 29%, which is low when compared with the latter. See Fig. 3.5.

The results representing a user with very little knowledge in URLs is very similar with approximately 38% correct when selecting unsafe with phishing URLs and a little less than 29% in selecting safe with safe URLs. The pre-population training with BERT doesn't seem to affect the classification in a relevant manner. See Fig. 3.6.

When using the safe URLs with the similar URLs we do see a great hit in accuracy as seen in Fig. 3.7 and Fig. 3.8 below. Fig. 3.7 shows BERT similarity with pre-population training of 30 safe URL and 15 similar URLs.

The accuracy is hit by having the second highest percentage when selecting safe on similar URLs. This means that the model struggled to differentiate between a safe URL and a phishing URL that is very similar to the safe, original one. Selecting not safe on a phishing URL is still the highest with less than 41%, but selecting safe on safe URLs is right below 17%. Selecting phishing on safe URLs is above 33%.

This model is not an accurate representation of how a person can analyze a URL. Since BERT is utilizing semantic similarity on the components of the hostname, it is not

Confusion Matrix
Actual Values
Phish
Safe
38.2
11.99
20.87
28.93
Phish
Safe
35
30
25
20
15

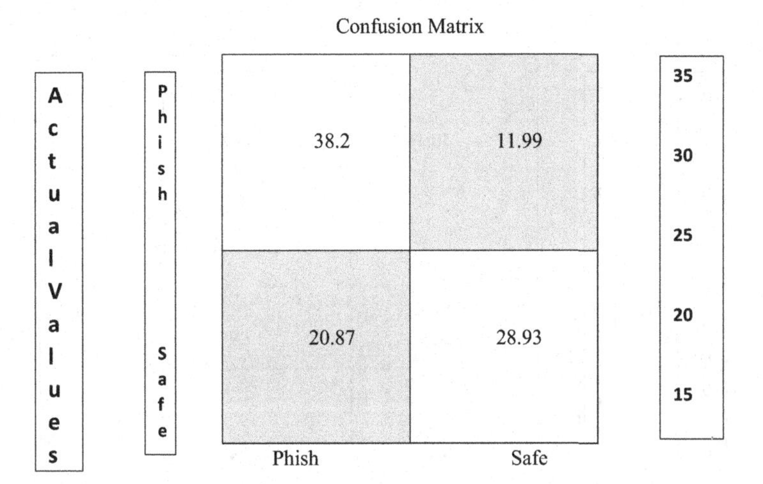

FIG. 3.5 BERT similarity with pre-population training of 30 safe URLs and 15 phishing URLs.

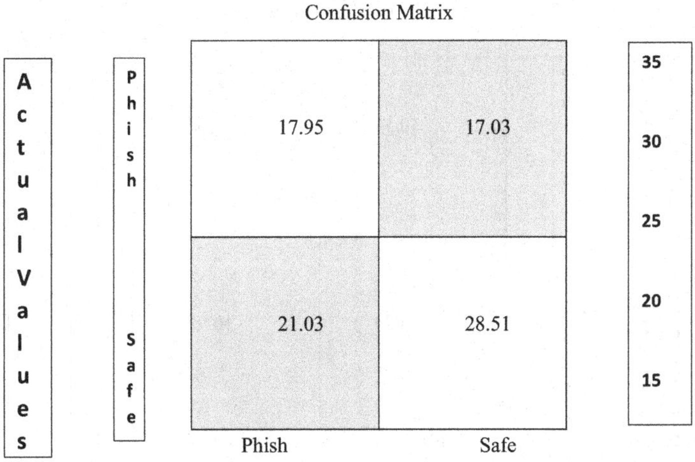

Confusion Matrix
Actual Values
Phish
Safe
17.95
17.03
21.03
28.51
Phish
Safe
35
30
25
20
15

FIG. 3.6 BERT similarity with pre-population training of 10 safe URL's and 0 phishing URLs.

accurately appointing the similarity of each word. The goal is to capture the accuracy to measure the improvement of selecting correctly when presenting a URL and BERT is more fitted to comparing documents and sentences thanks to its NLP prowess.

Running the custom similarity on the model showed different results as we can see in both Fig. 3.9 and Fig. 3.10, the accuracy is much more favorable to those presented with BERT's similarity. See also Fig. 3.11 and Fig.3.12.

As we can see, we get 69% accuracy in the most trained model while we get 66% accuracy in the least trained model. When dealing with URLs that are similar to those that are safe URLs, the model has a harder time differentiating between them. Custom similarity though, gets favorable results because of its lexical similarity, similar to

Confusion Matrix
Actual Values
Phish
Safe
40.16
9.49
32.31
17.97
Phish
Safe
35
30
25
20
15

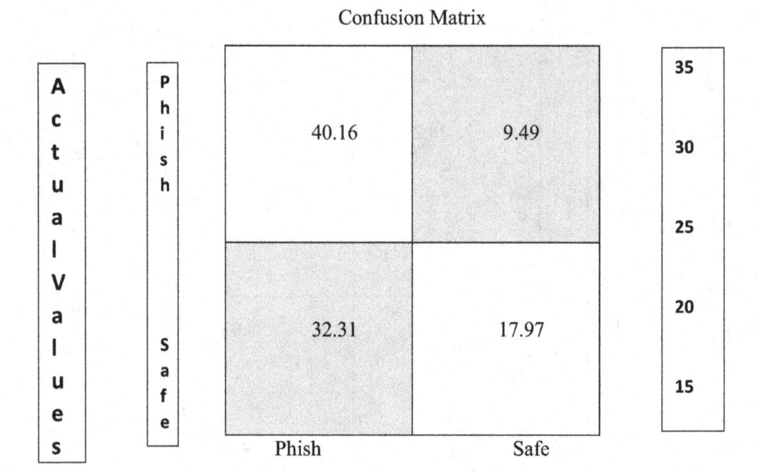

FIG. 3.7 BERT similarity with pre-population training of 30 safe URL's and 15 similar URLs.

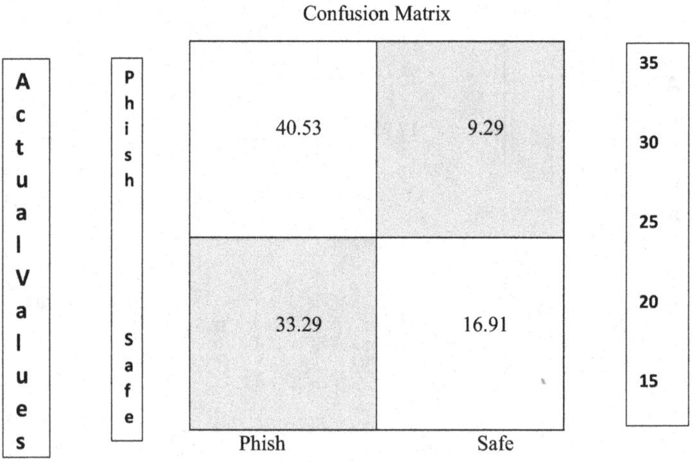

Confusion Matrix
Actual Values
Phish
Safe
40.53
9.29
33.29
16.91
Phish
Safe
35
30
25
20
15

FIG. 3.8 BERT similarity with pre-population training of 10 safe URL's and 0 similar URLs.

Jaccard's similarity, and that it is just taking the word of the URLs that it knows and comparing the letter by letter to the ones it doesn't.

6.2 Learning Results

The IBL model did best when utilizing the Basic Comprehensive Similarity than with BERT. This is clear because we don't get any benefit in semantic similarity as it may represent a way of overthinking and/or over analyzing. The model (BERT with pre- population training of 30 safe URLs and 15 phishing URLs, and 10 safe URLs with 0 phishing URLs) has a horizontal regression throughout when trained with the

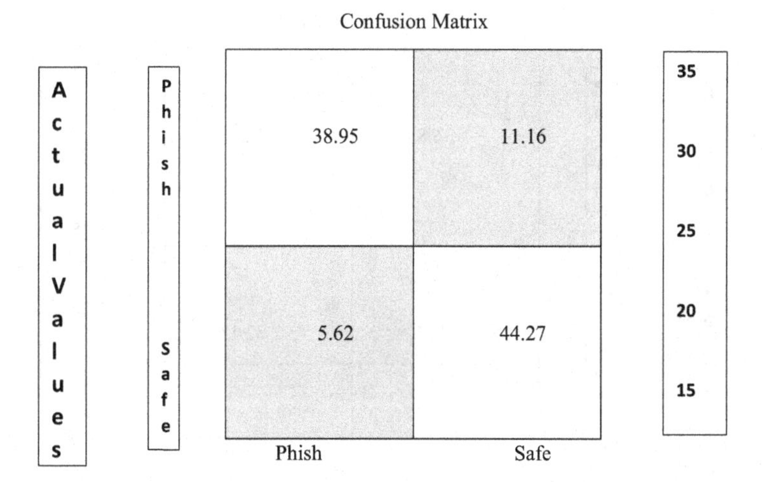

FIG. 3.9 Custom similarity with pre-population training of 30 safe URL's and 15 phishing URLs.

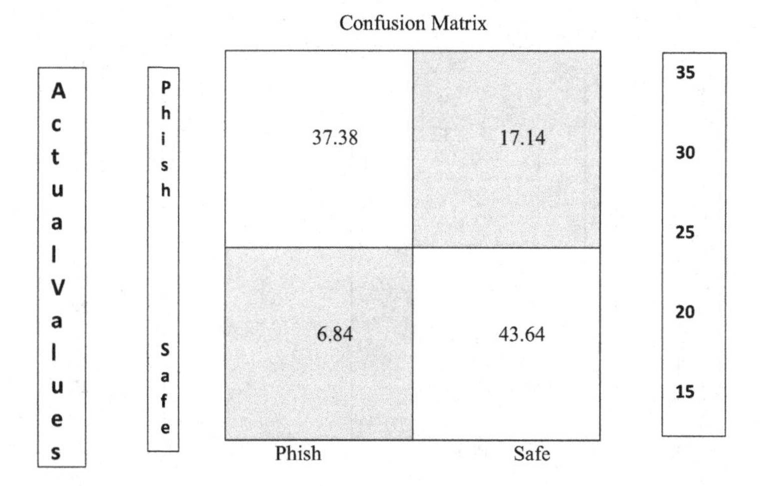

FIG. 3.10 Custom similarity with pre-population training of 10 safe URL's and 0 phishing URLs.

pre-population of 30 safe URLs and 15 phishing URLs and is very similar to the less trained model.

The same can be said when using safe URLs with similar URLs. We note in this case that participants don't improve as the trials continued.

When using the Basic Comprehensive Similarity, we can see a great deal of improvement in BERT with pre-population training of 30 safe URLs and 15 similar URLs, and 10 safe URLs with 0 similar URLs, as it shows that experience can make a difference as the trials increase.

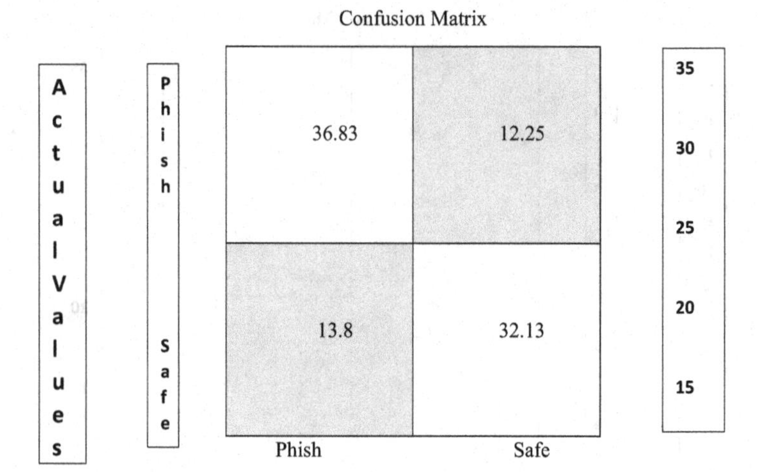

FIG. 3.11 Custom similarity with pre-population training of 30 safe URL's and 15 similar URLs.

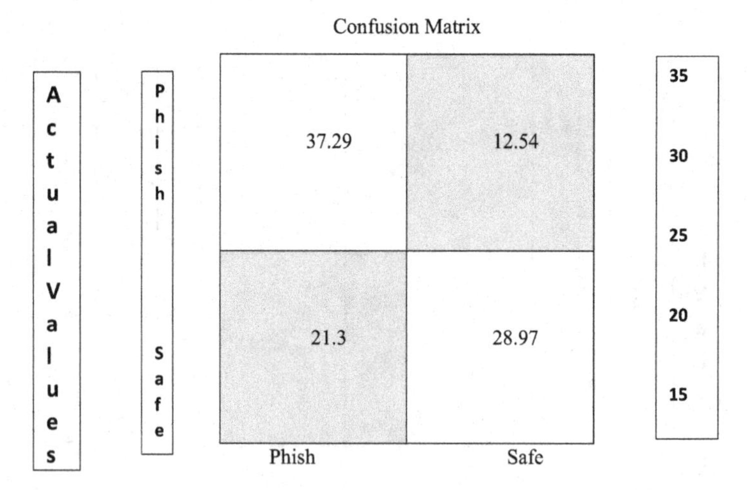

FIG. 3.12 Custom similarity with pre-population training of 10 safe URL's and 0 similar URLs.

Both results (Basic Comprehensive with pre-population of 30 safe URLs and 15 phishing URLs, and 10 safe URLs with 0 phishing URLs) show interesting findings in that the model can improve close to 80% regardless of the training.

With the Basic Comprehensive Similarity being used in the IBL model, we find that the increase is similar when using similar URLs as the malicious URLs when compared to the phishing ones. Custom with pre-population of 30 safe URLs and 15 similar URLs, and 10 safe URLs with 0 similar URLs; there isn't much difference as they both get close to 80% increase.

We can conclude that the IBL models using the Basic Comprehensive text similarity function show more prominent results than when it is using BERT. We can also conclude promising results to people that with basic training they can get close to 80% accuracy when deciding to click on a URL. This is good news as the majority of the URLs used for phishing are not similar to popular URLs.

There are occasions where similar looking to well-known URLs is used to deceive people. Based on the model, people may get over 70% chance of identifying whether a URL is malicious or not. This is good news as based on the model (showing a comparison to using BERT and the Basic Comprehensive Similarity with the 30 safe URLs and 15 phishing URLs) making sure to pay attention to the URL before clicking it will help to make a more informed decision and reduce the likelihood of falling victim to almost 80%.

7 DISCUSSION

The results provided prove that a person can make an analysis of the URL and make a favorable decision as to whether it is malicious or not based on the results of the IBL model. As experience is gained, the potential of making the wrong call falls to around 20%. The goal in mind is to demonstrate that phishing URLs can be analyzed and with a basic understanding of them can help identify potential threats.

Further work can be provided by adding relevance to the other components of the URLs to see if there are identifiable patterns that may help detect if they are malicious or not. There are some limitations that are important to consider since browsers don't need the subdomain in the URL for them to work. Additional work can also be done by working in conjunction with current technologies to determine the statistical probability of falling victim to phishing and analyze such URLs as they might show what future tactics will be used by malicious actors. Another aspect is to initiate human trials to compare human behavior against the IBL model presented. This can show the effectiveness of the model and where future research may be needed.

REFERENCES

[1] Statista. (2022). Global digital population as of April 2022. *Works*. Transactions of the Association for Computational Linguistics, 8, 842–866. www.statista.com/statistics/617136/digital-population-worldwide/

[2] Cranford, E. A., Rajivan, P. & Aggarwal, P. (2019). *Modeling Cognitive Dynamics in End-User Response to Phishing Emails*. ICCM-Conference. https://iccm-conference.neocities.org/2019/proceedings/papers/ICCM2019_paper_57.pdf

[3] Ubing, A. A., Kamilia, S., Abdullah, A., Jhanjhi, N. Z. & Supramaniam, M. (2019). Phishing website detection: An improved accuracy through feature selection and ensemble learning. *International Journal of Advanced Computer Science and Applications*, 10(1), 252–257. https://doi.org/10.14569/ijacsa.2019.0100133

[4] KnowBe4. (2017). History of phishing. *Phishing.org*. Retrieved May 9, 2022. www.phishing.org/history-of-phishing

[5] Shonman, M., Li, X., Zhang, H. & Dahbura, A. (2018). *Simulating Phishing Email Processing with Instance-Based Learning and Cognitive Chunk Activation*, pp. 1–11.

[6] Xu, T., Singh, K. & Rajivan, P. (2022). Modeling phishing be analyzed and with a basic understanding of decisions using Instance Based Learning and Natural Language Processing in *Proceedings of the Annual Hawaii International Conference on System Sciences*, pp.1–10.

[7] James, J., Sandhya, L. & Thomas, S. J. L. (2013). Detection of Phishing URLs to see if there are identifiable patterns that may use machine learning techniques in *International Conference on Control Communication and Computing (ICCC)*, pp. 303-309.

[8] Semrush. (n.d.). Top 100: The most visited websites in the US [2021 top websites edition]. Retrieved May 9, 2022. www.semrush.com/blog/most-visited-websites/

[9] PhishTank. (n.d.). PhishTank > see all suspected phish submissions. *PhishTank.* Retrieved March 17, 2022. https://phishtank.org/phish_archive.php

[10] Lejarraga, T., Dutt, V. & Gonzalez, C. (2012). Instance-based learning: A general model of repeated binary choice. *Journal of Behavioral Decision Making*, 25(2),143–153.

[11] Carnegie Mellon University. (n.d.). *PyIBL 4.2 documentation*. Retrieved February 7, 2022. http://pyibl.ddmlab.com/

[12] Rogers, A., Kovaleva, O. & Rumshisky, A. (2020). A primer in Bertology: What we know about how Bert works. *Transactions of the Association for Computational Linguistics,* 8, 842-866. https://doi.org/10.1162/tacl_a_00349

[13] Briggs, J. (2021, September 2). Bert for measuring text similarity. *Medium*. Retrieved May 9, 2022. https://towardsdatascience.com/bert-for-measuring-text-similarity-eec91c6bf9e1

[14] Alshira'H, M. & Al-Fawa'reh M. (2020). Detecting Phishing URLs using Machine Learning Lexical Feature-based Analysis, *International Journal of Advanced Trends in Computer Science and Engineering.* ISSN 2278-3091, 9(4), July–August 2020. www.warse.org/IJATCSE/static/pdf/file/ijatcse242942020.pdf, and https://doi.org/10.30534/ijatcse/2020/242942020

Section II

Cybersecurity Threats to the Individual

4 Video Games in Digital Forensics

Heba Saleous and Marton Gergely
College of Information Technology, United Arab Emirates (UAE) University
Al Ain, United Arab Emirates

1 INTRODUCTION

The digital world has been booming with activities previously unforeseen by users and researchers alike. What was once a platform for professional communication and research has now become a necessity in the daily lives of people around the world. Technology and the Internet are now being used to communicate with friends and relatives from across the globe, offer distance learning and work during the COVID-19 pandemic, and stream live entertainment. One form of entertainment that has been rising in popularity is video games. This is due to its evolution into a much larger industry and the advancement of computer hardware. The video game industry, in addition to continuing its role in providing entertainment, has introduced new careers and been a factor in the increase of network communications.

As a result, video gaming has greatly increased in popularity over the years. In addition to the increased number of gamers around the world, the esports and streaming populations have also grown. This popularity has brought to light the many benefits of video games. Charity organizations have begun to sponsor and host streams as a means to gather more money and spread awareness about the issues being targeted. Video games also have benefits in both physical and mental healthcare; patients can work on their weight and body strength with active video games or find stress relief in some mainstream games. The active benefits of video

DOI: 10.1201/9781003415060-6

games can also be used in physical education in schools to encourage students to exercise. Additionally, games can also be used to improve teaching in other subjects, such as science or math.

Despite all of these benefits, however, the rise in popularity of video games also raises some concerns. Adversaries can use video games as a whole new platform for their malicious activities, which include covert communication, disruptions in service, or remote code executions. These new threats have raised questions about the extent to which video games, especially consoles, can be used for cybercrimes. The differences in systems between a standard personal computer (PC) and a video game console may pose challenges for investigators, potentially hindering investigation progress. Video games and consoles will need to be considered more in digital forensics research to overcome these challenges and prevent time- and resource-consuming obstacles in the future.

1.1 Rising Popularity

Over the years, the popularity of video games has risen and is continuing to rise. This is especially noticeable during the COVID-19 pandemic. According to USA Today, there was a spike in the number of gamers between the years 2019 and 2020 in the United States alone, approximately 70 million players (Snider, 2021). Worldwide, there are an estimated 2.7–2.8 billion gamers over various devices and platforms around the world as of 2021 (Kelly et al., 2021; Wijman, 2020). This growth in the number of people playing video games is a trend that has been noted for several years now. The monetary value of the gaming industry has also been increasing, becoming more valuable each year that passes and evolving into a multi-billion-dollar market. The market value of the video game industry almost surpassed $180 billion in 2021, up from $156 billion in 2020 (Video Game Industry Statistics, Trends and Data In 2022; Wijman, 2021; Williams, 2022).

In addition to the increase in the number of players, some multiplayer video games are now being considered sports, dubbed esports, such as League of Legends, DoTA2, Counterstrike: Global Offensive, and Overwatch. Esports games add a competitive level to the entertainment provided by video games, providing more opportunities for players wanting to participate. There are two types of esports players: those that join professional teams and play in world championships, and those that stream their gameplay. Over the years, the professional side of esports has grown. What was once just a few teams competing for a few thousand-dollar prize pool has evolved into a billion-dollar market. As of 2021, the esports market was estimated to be almost $1.3 billion (Hagan, 2021; Kelly et al., 2021; Wijman, 2021).

Streaming platforms, such as Twitch and YouTube, have become their own markets. The number of both streamers and viewers has been increasing steadily over the past few years, especially during the COVID-19 pandemic. Viewership on Twitch has spiked since 2019. Between 2019 and 2020, there was a 67.36% increase in the total number of hours streamed, from 11 billion hours to 18.41 billion hours (Hagan, 2021; May, 2021). Between the same time period, the average number of concurrent users jumped from 1.3 million viewers in 2019 to 2.1 million in 2020 (May, 2021; Williams, 2022). While streaming is initially intended for video games, these platforms have transformed into sharing platforms for not only gaming communities, but also for music, art, or general socialization. These streams have clearly had a big impact on players.

1.2 Rising Benefits

Video games potentially have various beneficial traits that can be taken advantage of in different ways. The most common benefits of video game incorporation are to raise money for charities, improve mental and physical wellbeing, and assist with teaching and student motivation. A few notable charities to name are AbleGamers (AbleGamers, n.d.), The Red Cross (Red Cross, n.d.), Help for Heroes (Help for Heroes, n.d.), Extra Life (Extra Life, n.d.), Games Aid (Games Aid, n.d.), St. Jude's Children's Hospital (St. Jude's Children's Research Hospital, n.d.), and dedicated streams by content creators on Twitch (Twitch, n.d.). Additionally, some third-party video game sellers, such as Humble Bundle (Humble Bundle, n.d.) and Groupees (Groupees, n.d.) allow customers to donate a part of their payment to the charity of their choice. The various benefits of video games are summarized in Table 4.1.

1.3 Rising Threats

Multiplayer games have greatly grown in popularity, providing a whole new platform for malicious activity. Historically, video games have been used by cybercriminals for a variety of nefarious activities that vary in complexity and motive, whether it is ruining the gaming experience for fellow players out of sheer enjoyment or compromising a company's servers to steal private information. Table 4.2 lists some potential threats of video games with examples of the relevant attacks that have occurred in the real world.

1.4 Problem

Video games have clearly become a popular, major part of our lives, evolving from being a simple pastime to becoming a major industry. Criminals around the world have begun to see the importance of gaming and are taking advantage of the new opportunities for attack and covert communications that come with it. The extent of this importance has reached a point where, during the terrorist attack in France in 2015, spotting a PlayStation 4 during one of the raids in the hunt for the culprits sparked many questions and raised concerns about how video games can be used for malice (Tassi, 2015). The uncertainty of how to approach the problem at the time, and the level of alarm raised at the possibility, made it clear that video games need to be considered more during incidents, and investigators must be better prepared for such encounters.

TABLE 4.1
Beneficial Aspects of Video Games

Benefit	Description	Examples	References
Charities	Video game streams can allow viewers to donate money, which can be used either for their own livelihood or to donate to a sponsored charity.	1. Stream platforms, such as Twitch, have adopted systems dedicated to charity streaming. 2. Well-known charity organizations, such as the Red Cross, Help for Heroes, and St. Jude's Children's Research Hospital, collaborate with streaming platforms to host charity-dedicated streams. 3. Some video game resellers, such as Humble Bundle or Groupees, give some of the proceeds from purchased games to various charities. 4. Charities that revolve around video gaming, such as AbleGamers Charity, Games Aid, and Extra Life, have been established.	(AbleGamers, n.d.; American Red Cross, n.d.; Extra Life, n.d.; Games Aid, n.d.; Groupees, n.d.; Help for Heroes, n.d.; Humble Bundle, n.d.; St. Jude's Children's Research Hospital, n.d.; Twitch, n.d.)
Mental Healthcare	Video games can be incorporated into mental health treatment due to the psychotherapeutic effect some have on patients. Additionally, they improve cognition and concentration by incentivizing treatment goals, rewarding players for activities, and promoting behavioral change.	1. The game Rayman was found to help reduce the levels of various types of anxiety. 2. Treatments for mood disorders that incorporate video games are found to be non-invasive, do not have the negative side effects of medicinal treatments, and have a better cost-effectiveness ratio. 3. The Food and Drug Administration (FDA) approved the game EndeavorRX for ADHD treatment.	(Boldi and Rapp, 2021; Hollister, 2020; Kowal et al., 2021; Peñuelas-Calvo et al., 2022)
Physical Healthcare	Video game can be used for active physical exercises to manage body weight and improve overall physical fitness.	1. Dance, Dance, Revolution, Wii Fit, and Ring Fit Adventure can help with weight loss and physical therapy by encouraging the player to complete physical activities to progress through the game.	(Comeras-Chueca et al., 2021; Oliveira et al., 2020; Santos et al., 2021)
Education	Video games can be used in a variety of school and university subjects to improve teaching methods and enhance student motivation through incentivization.	1. Video games have been introduced in physical education (PE) in schools, with evidence that physical activity and fitness were improved. 2. Games can assist with and improve learning and motivation with subjects, such as languages, vocabulary, sciences, history, and mathematics.	(Comeras-Chueca et al., 2021; Gordillo et al., 2022; Lopez-Fernandez et al., 2021; Martinez et al., 2022)

TABLE 4.2
The Cyber Threats That Have Occurred via Online Video Games

Threat	Description	Examples	References
Covert or Terroristic Communication	Covertly communicating with allies through the game or game-related channels. Malicious users may also abuse this to radicalize and recruit members into extremist or terrorist groups.	1. Teenagers in the United Kingdom were found to be radicalized by white supremacy and neo-nazi groups to perform acts of terrorism. 2. Extremists have used games, such as Minecraft, Call of Duty, and World of Warcraft (WoW), and encrypted communication channels, such as Telegram or Discord, to spread hate and radicalize players. 3. American and British spies have used WoW and Second Life to conduct surveillance and communicate.	(Mazzetti and Elliot, 2013; Miller and Silva, 2021; Townsend, 2021)
Denial-of-Service (DoS)	Denying players services by forcibly removing them from online games through the use of exploits or attacking game servers.	1. Titanfall and Titanfall 2 were targets for DoS attacks by begrudged players, rendering the games unplayable for others and resulting in discontinued support from the developers. 2. Various major companies, such as Riot Games, Valve, and Blizzard, fell victim to the infamous "DerpTrolling", who sought to take down their servers and deny players around the world their services simply for their own enjoyment.	(Chalk, 2021; Cluely, 2019; Robertson, 2018; Winkie, 2021)
Remote Code Execution (RCE)	The ability to remotely run malicious scripts that affect other players in online game modes.	1. Bandai Namco, the publishers of the Dark Souls series and Elden Ring, had to take down their Player-versus-Player servers to investigate an issue that allowed malicious players to hijack their opponent's PC through RCE. 2. Players of Amazon's mass multiplayer online (MMO) game, New World, discovered that the game was released with an unsecure chat box and abused it to inject HTML code for purposes ranging from posting oversized images to inconvenience others to DoS attempts.	(Abrams, 2016; "Archimtiros", 2021; Birnbaum, 2016; Grustniy, 2021; Kim, 2021; Litchfield, 2022; Roth, 2022)

(Continued)

TABLE 4.2 (Continued)
The cyber threats that have occurred via online video games

Threat	Description	Examples	References
		3. WoW's ability to install addons, while relatively safe, sometimes result in malicious scripts. More specifically, the WeakAuras addon, which allows players to not only write scripts but also share them, was victim to the sharing of a malicious script that overpriced auction house items once the transaction was confirmed. 4. WoW also experienced RCEs through its chat feature. Malicious players social engineered others into typing scripts into the chatbox that resulted in having their accounts hijacked in-game.	
Trojanized Third-Party Applications	Third-party applications can be installed to manage addons or provide in-game overlays that improve gameplay. However, malicious users can take advantage of players' dependency on these applications by creating identical trojans that perform malicious activities in the background.	1. In 2014, a trojanized version of the Curse client, used to download addons for popular games such as WoW, Rift, Minecraft, and The Elder Scrolls Online, was made available online as an update that, when installed, would bypass authentication to steal user account information.	(Goodin, 2014; Karmali, 2014)
User Data Theft	Malicious parties can attack a publisher's servers to gather sensitive data.	1. Electronic Arts, Capcom, and CD Projekt Red recently fell victim to attacks on their servers that were confirmed to have collected private user data, such as credit card numbers and home addresses.	(Criddle, 2021; Greig, 2022; Hamilton, 2020; Purslow, 2022; Ray, 2021; Tidy, 2020)

1.5 Objective

The objective of this work is to provide insight on existing literature related specifically to both video games and digital forensics. Given the current state of video games and the constant evolution of technology, it is imperative to address the challenges being presented. In this chapter, a literature review specifically targeting video games in the field of digital forensics is conducted. The included literature is analyzed and categorized to determine the work that has been completed so far, and where the research gaps remain.

The remainder of this chapter is organized as follows: Section 2 describes the methodology followed in this work and the actual review of the literature. Section 3 provides an analysis of the literature included. Section 4 discusses the remaining challenges to be addressed before concluding the paper in Section 5.

2 LITERATURE REVIEW

2.1 Methodology

Video games have clearly evolved over the years and embedded themselves into today's society. As a result, there is a clear need to take these games, and the devices with which they are played, into consideration during investigations. To begin addressing the newly presented challenges, the following questions need to be asked:

1. What research has already been done related to video games in the field of digital forensics?
2. What more can be done to improve on existing research and advance the work done related to these two topics?

To begin answering these questions, a targeted literature review is conducted using the following search query:

((Digital || Computer || Cyber) && (Forensics || Investigations)) && (Games || Gaming)

This query was submitted to major databases, such as Elsevier, IEEE, ACM, and Springer. Google Scholar was also used to find publications that may not have been published to any of the listed databases. As a result, 56 publications were found related to both video games and digital forensics. After being reviewed, these works were categorized in several ways to help determine the overall focus of the paper. The following are the categories used:

1) **Device**: The device of focus is recorded to filter out the environment that the authors conducted their research in.
2) **Connectivity**: Since video games can be played either offline alone or online with other players, the connectivity of the research conducted is recorded.
3) **Aspect**: This category defines whether the authors focused on physical or virtual properties of the system being investigated.

   a. **Physical**: The physical aspect includes the physical properties of a device, such as raw memory dumps or any tangible physical modifications.
   b. **Virtual**: The virtual aspect features intangible, software-related features of a system, such as gameplay exploits or simulated environments.
4) **Paper Type**: Recording the type of paper being reviewed is helpful for filtering the overall content included.
   a. **Framework**: Papers that present frameworks or models with no proof-of-concept are categorized here.
   b. **Implementation**: In these papers, the authors presented both a model and an implementation of their proposed solution.
   c. **Review**: Publications categorized as reviews provide some insight on existing work done with video games in the context of digital forensics.
   d. **Guide**: Works that provided a detailed step-by-step process that the authors followed to research a solution are classified as guides.
   e. **Survey**: The study revolved around a questionnaire that was passed onto a specified population to gather information about the subject at hand.

In addition to the categories described above, keywords were assigned to each publication that assisted in summarizing the content within. These keywords will be used to summarize the papers in the following sub-sections.

2.2 Behavior Analysis

With digital forensics being considered a part of both law enforcement and cybersecurity, it is important to take into consideration the behavior patterns of criminals. Doing so would help determine motive and objective, as well as predict potential actions. In this review, three of the 56 publications focused on behavior analysis with regards to video games and digital forensics. Each of these works were done on computers, focused on online interactions, and analyzed an aspect of Massively Multiplayer Online (MMO) games.

Kwon et al. (2016) focused on proposing a framework to detect gold farming groups in AION, a popular MMO. Gold farming can be harmful to the game's economy by hiking prices and limiting resources and equipment available in in-game markets. Ahmad et al. (2011) focused on a similar aspect of MMOs by analyzing and profiling malicious transactions that occur in such markets. Steering away from virtual markets, Drachen et al. (2012), gathered data from two games, Tera and Battlefield: Bad Company 2, to create behavior profiles for players and their characters within the games.

2.3 Cybercrime

In addition to video games becoming a source of entertainment and a means to socialize with others around the world, criminals have found that they also offer more opportunities for malicious acts. Several authors sought to study some of the malicious activities that have been found in video games, both online and offline.

West (2015) and Rakitianskaia et al. (2011) reviewed the potential cybercrimes that can be committed through video games, such as fraud and identity theft. Additionally, they make various suggestions for how to address and investigate the crimes that occur.

Podhradsky et al. (2011b) attempt to look at cybercrimes from the other side; from the perspective of the criminal rather than the investigator. They attempt to acquire personal information from old, discarded consoles. Based on their experiments, the authors offer suggestions for how to properly sanitize and dispose of unwanted video game consoles.

The final publication that was assigned to this keyword was written by Wijaya et al. (2014). They provided a framework for a potential solution to the cybercrimes that occur through video games, namely fraud. The authors suggested including phone numbers and personal identification verification, as opposed to the standard email address, to create and register accounts with online gaming platforms.

2.4 Education

As technology continues to evolve and become increasingly important in the daily lives of people around the world, the true importance of education and training in cybersecurity begins to come to light, especially when focusing on digital forensics. There is a known shortage of digital forensics specialists around the world, emphasizing the importance of improved education in this field. Some authors sought to address this challenge by incorporating digital forensics education with video games due to the rising popularity of gaming. Several publications in this review (Blauw and Leung, 2018; Blazic et al., 2016; Crellin and Karatzouni, 2009; Jin et al., 2018; Nordhaug, 2014; Pan et al., 2015, 2012; Yerby et al., 2014) focused on using video games to improve digital forensics education for students. Games that allowed students to be put into the perspective of the investigator were developed and tested. To progress with the lesson and learn about the proper investigative process, certain tasks must be completed.

The increased presence of video games and the devices and hardware used to play them means that they are becoming more common at the scene of a crime. Therefore, it is imperative for investigators to consider these devices and learn how to analyze them. Other publications that were assigned the "education" keywork (Conway et al., 2015; Drakou and Lanitis, 2016; Herr and Allen, 2015; Karabiyik et al., 2019) focused on developing serious games for active investigators in law enforcement or the Department of Defense, aiming to train them in how to handle video game-related objects that may be found while investigating an incident. However, the training focuses on studying devices found at the scene of the incident rather than the software involved, leaving a gap in research regarding this topic of study.

2.5 Forensic Ability

The rise in popularity of video game consoles combined with the previously unconsidered potential malicious uses of these devices poses the question if modern forensic methods and tools are even usable in investigations. Some researchers saw this question as an opportunity to explore this possibility, selecting a console and trying to follow standard investigative protocol to forensically acquire and analyze it. The results of their work were used to create guides for other investigators, explaining the exact process followed to ensure that data is forensically sound and uncorrupted.

Conrad et al. (2009, 2010), Davies et al. (2015), and Khanji et al. (2016) explored video game consoles created by Sony, namely the PlayStation 3, PlayStation 4, and PlayStation Portable. Many other authors (Al-Haj, 2019; bin Mohd Isa, 2009; Burke and Craiger, 2007a, 2007b; Cravel, 2015; Elfving and Lidstrom, 2020; Khanji et al., 2016; Luttenberger et al., 2012; Moore et al., 2014; Podhradsky et al., 2011a, 2011b, 2012, 2013; Xynos et al., 2010) chose to investigate Microsoft's consoles, the Xbox, Xbox 360, and Xbox One. The number of authors discovered for Microsoft's consoles exceed the number of Sony's, likely due to the fact that the operating system found in the Xbox is similar to Microsoft Windows but modified specifically for the console, whereas the PlayStation's operating system is entirely unique and requires more effort to access and understand. Few authors (Barr-Smith et al., 2021; Pessolano et al., 2019; Read et al., 2016; Turnbull, 2008) attempted to investigate Nintendo consoles, more specifically the Wii, 3DS and Switch. While these consoles are popular, Nintendo has a reputation for making them extremely difficult to modify.

In addition to exploring the forensic ability of investigating video game consoles, some authors turned their attention towards specific games or other virtual aspects. Games with online communications can be taken advantage of for malicious reasons. Some authors (Ebrahimi and Lei Chen, 2014; Morris, 2012; Rakitianskaia et al., 2011; Tabuyo-Benito et al., 2019; Taylor et al., 2019; van Voorst et al., 2015; Yarramreddy et al., 2018) investigated these games to determine what kind of artifacts can be acquired, creating a guide in the process to provide investigators with a reference in the event they encounter such platforms. Others (Gibbs and Shashidhar, 2015; Hale et al., 2012; Irmler and Creutzburg, 2014) focused on how criminals can take advantage of the more in-depth, virtual aspects of video games, such as game engines and maps.

2.6 Legal

Researchers have explored the potential use of video games in the legal system to improve evidence representation in the court room. Schofield (2009, 2011) focused on video game graphics for forensic animations and evidence representation. They believed that doing so can help courtroom attendees better understand the case being observed and give a more educated judgement.

2.7 Privacy

With the increased network presence of video games, privacy during online interactions starts to become a larger concern. Two publications in this review targeted privacy concerns within online gaming. Fuchs (2012) explored online surveillance by game companies while discussing privacy challenges with video games. Podhradksy et al.

(2013), on the other hand, elaborated on both privacy and security flaws found within Xbox's online services and file system, providing a guide on how they were able to break both.

2.8 Security

Security has become a rising concern worldwide with the increased presence of technology. Adding video games to the mix only makes this issue more complicated with the unorthodox systems being introduced. Some authors turned their focus to these rising security concerns. Vaughan (2004) realized this issue early on, exploring the security concerns found within the first-generation Xbox console. Chen et al. (2016) also explored this issue, on concerns regarding some popular PC games and with the PS4 system.

With the rise in popularity of Blockchain environments, some game developers have created games that run on the Blockchain in hopes of increasing security or providing players with "play-to-earn" opportunities. Munir and Baig (2019) explored this type of gameplay model by creating their own Blockchain game to gain a firsthand perspective of the challenges that are encountered with these types of games.

2.9 Virtual Reality

Virtual Reality (VR) is a technology that is becoming increasingly popular. It has a variety of purposes beyond gaming, such as art, therapy, or education. In the context of digital forensics, researchers saw VR as a potential platform for improved education and training. Conway et al. (2015), Drakou and Lanitis (2016), Ebert et al. (2014), and Karabiyik et al. (2019) all created frameworks for VR use in digital forensics. Each of these publications described the process of creating a virtual environment that allowed investigators to reconstruct and visualize the scene of the crime while also providing them with the opportunity to train themselves in the investigative process.

3 ANALYSIS

The keywords used to categorize the literature review in the preceding section helped summarize and categorize each publication. However, there are other labels to consider in this study, such as the type of paper, the device studied, online or offline connectivity, and the layer of focus.

There were five paper types found during the review: Framework, Implementation, Review, Guide, and Survey. Frameworks discussed the issue being focused while providing a model for a potential solution. However, these works did not provide any proof-of-concept experiments and were theoretical. Papers labeled as Implementation provided both a model and an experiment to address an issue, discussing the results gathered. Review papers focused on an area of concern related to both video games and digital forensics, citing relevant challenges and suggestions found in other publications. Guides provided a step-by-step process for addressing the topic being discussed, sharing insights on challenges, shortcuts, and solutions discovered along

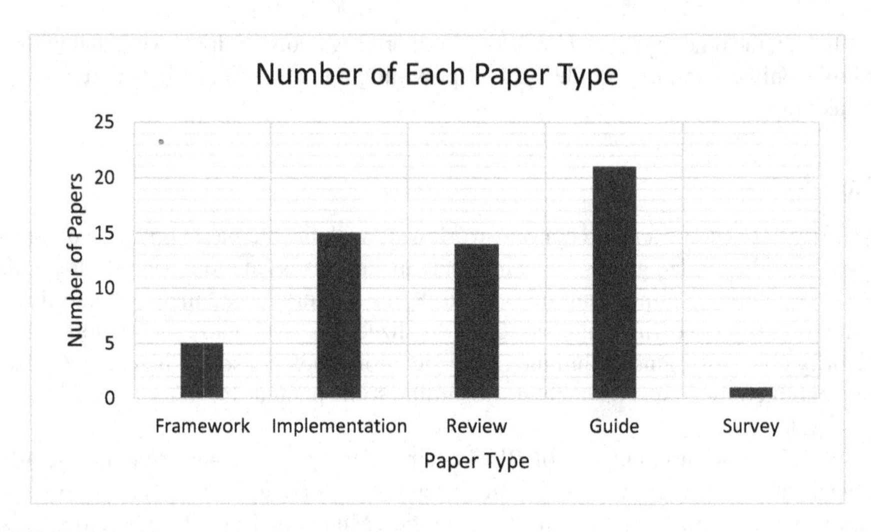

FIGURE 4.1 The number of papers found in each type.

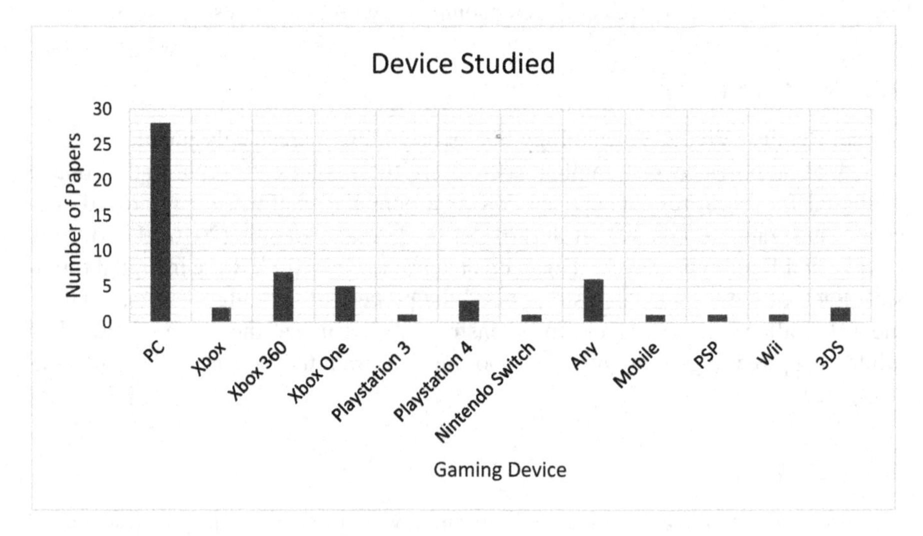

FIGURE 4.2 The number of papers for each device.

the way. Finally, Surveys reviewed a topic related to both digital forensics and video games by sending a questionnaire targeting a specific group of people. The number of each type of paper included in this review can be seen in Figure 4.1. As noted by the figure, most of the papers included in this review are Guides, followed by Implementations and Reviews.

The consideration of video games in the context of digital forensics means that there are many more devices to be focused on during an investigation, namely those that are used to actually play the games. The number of each device found in this study can be seen in Figure 4.2.

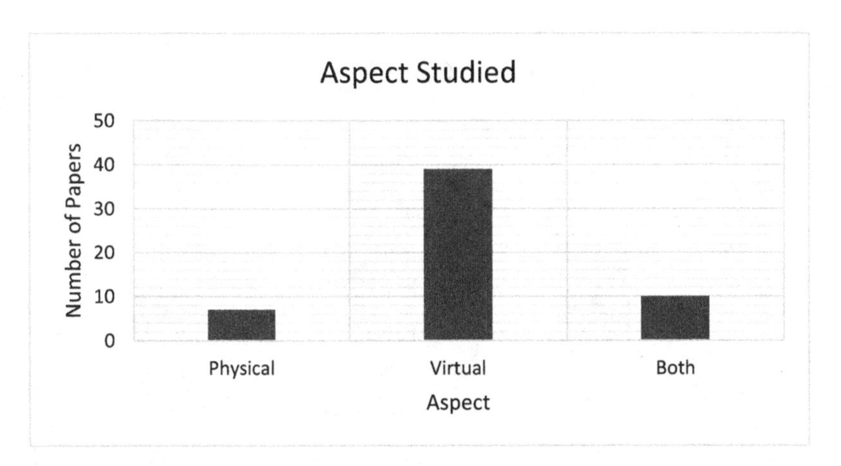

FIGURE 4.3 The number of papers in each label.

As seen by the graph, there was a significantly larger number of studies related to PC gaming compared to the other platforms. This emphasizes the stated concern of needing more focus on video game consoles in digital forensics research; while PC games pose just as much of a risk as console games, to an extent, they can be analyzed in a similar manner as standard PC executables. Additionally, they may leave behind artifacts, resulting from network communication, user data, and file or permission modifications, that can be studied using standard forensic tools. Console games, on the other hand, may be more difficult to investigate due to the less familiar operating systems that may be difficult to analyze with the available tools.

In addition to determining the type of paper and the devices involved in the research, the works included in this review were also categorized according to which aspect of the device was studied. In other words, papers were labeled based on whether they focused on the *physical* aspects of analysis, such as hardware modifications or digging into physical memory, or *virtual* characteristics, such as software tools or system features. The number of publications categorized with these labels can be seen in Figure 4.3.

Most of the works included in this study focused on virtual aspects of video games and their respective systems. However, some authors studied both physical and virtual aspects to try and provide a comprehensive study of the topic they are writing about. The lack of attention towards the physical aspect offers an opportunity for further research in this area; looking towards modifications to the hardware of a console that may assist adversaries with malicious activities. Considering the difficulty in analyzing console hard drives using standard forensic tools, physical modifications to a console, internal or external, may be difficult to study. One example of this would be a USB device that, when plugged into the console, modifies the functioning of the console, either by bypassing security measures or introducing another OS.

The final category taken into consideration in this review is the device connectivity. Along with the increased presence of networked devices all around us, gaming has also evolved to include more internet-based features. As such, the connectivity of

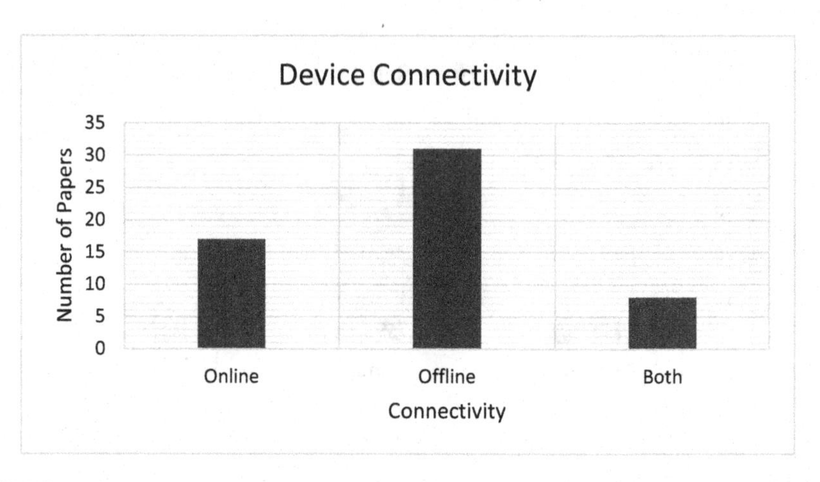

FIGURE 4.4 The number of papers that targeted online or offline features.

the topic being studied must be considered. Figure 4.4 depicts the number of papers included in this review that focused on offline or online features.

In most of the papers, the research conducted focused on offline features of video gaming with regards to digital forensics. These include models for improved education, guides for analyzing a console, or inspecting storage or memory devices. Some studies targeted online communications, such as inspecting browsing and communication history, or potential security and privacy concerns. Finally, a few of the publications in this review opted to research both online and offline features of gaming in the context of digital forensics. The rise in popularity of online multiplayer games means that devices are exposed to more network activity, resulting in more potential artifacts to explore. If studied, these artifacts can be used to map out other entities that the device may have interacted with, both voluntarily and involuntarily.

The number of devices found in each keyword used to categorize the literature in this review is summed up in Figure 4.5. The numbers depicted in this graph complement the claims stated earlier in this paper about the majority of publications focusing on PC-related topics. Revisualizing this gap in this manner further emphasizes the need to give other devices, namely video game consoles, more research attention. One other fact to note is that, based on the research done in this paper, there has been more work done on Microsoft's consoles (Xbox, Xbox 360, and Xbox One) than other consoles. This is likely due to two things. Firstly, Microsoft's consoles are popular among gamers because of big-name titles, such as the Halo, Fable, and Gears of War series, and secondly, likely because of the similarity between the Xbox consoles' and Window's file systems.

As stated earlier, the number of papers that are labeled as guides greatly exceeds those of other types. Figure 4.6 further supports this statement by showing how many papers of each keyword in this review belonged to each paper type. Almost all papers listed under the "forensic ability" keyword were classified as guides, which are papers that describe the steps taken by the authors to try and analyze the device or feature in question. However, the clear gap among the other paper types when looking at this

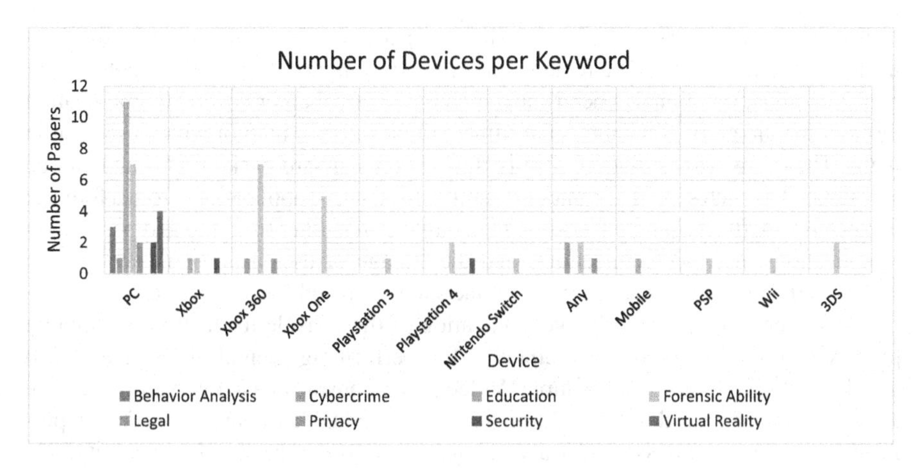

FIGURE 4.5 The number of devices for each keyword.

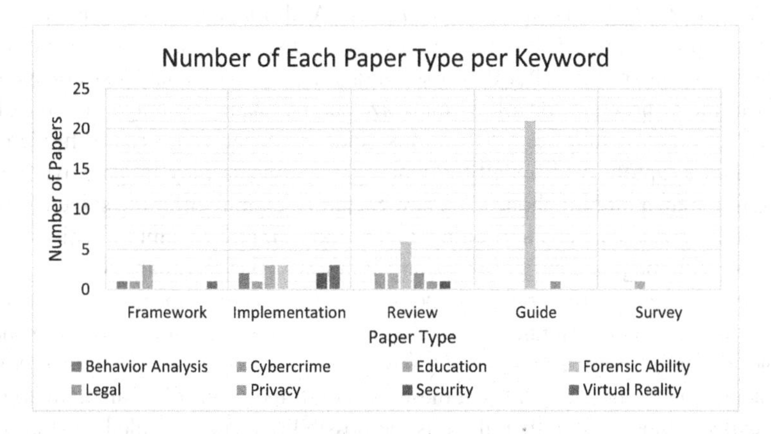

FIGURE 4.6 The number of papers belonging to each paper type for each keyword.

keyword indicates that further research will need to be done to expand our knowledge in how video games and consoles affect digital forensics. Publications related to "education" were found in every other paper type due to the various ways authors explored video games in digital forensics with respect to educating professionals and improving existing training methods.

4 DISCUSSION

Previous research has focused on a variety of topics related to how video games affect digital forensics research. More notably, the papers included in this review cover how video games can be used to improve digital forensics training and education, guides for how past generation video game consoles can be forensically studied with current tools, and potential security, privacy, and cybercrime concerns. Based on our findings, most of the papers included in this literature review focused either on creating forensic analysis guides for video game consoles or on PC-related topics and concerns.

One concern that was not covered by these papers, however, is the potential weaponization of these consoles. The review did include papers that discussed the potential to perform online, traditional cybercrimes, such as identity theft or covert communication through steganography in games like Minecraft or Terraria, both of which use custom maps. These are crimes that are also performed on standard desktop PCs and can be investigated as such since PC games are similar to standard applications from a forensic perspective. Artifacts can be found on the computer resulting from file and permission modifications, network connections, and the storage of user data. Standard forensic suites should be able to analyze these artifacts to gain insight on the case at hand.

The concern arises with the weaponization of the console itself. PCs and mobile phones can be weaponized specifically for performing Denial-of-Service (DoS) attacks, or surveillance, for example. Video game consoles have evolved to become almost as powerful as desktop PCs, providing near-equal hardware and software performance. It may be possible for a video game console to have not some, but all of the malicious capabilities of a PC. The difference in operating and file systems, such as with the level of security on the console or the configuration of the software stack, create new challenges that must be addressed. Additional concerns arise from how much more portable and covert some consoles are than laptops. The Nintendo Switch or 3DS is significantly smaller than the average laptop and is also less conspicuous. One example of how the console's small size can be taken advantage of is the ability to sneak the device into secure locations since it can be more easily hidden on a person than a laptop. Additionally, a video game console may be perceived as less threatening than a mobile phone. The new Steam Deck may offer the same opportunities as these consoles, despite running a more familiar OS (SteamOS, which is based on Linux, with the possibility of installing Windows 10 (Freedman, 2022)).

There is a need to further research the possibility of video game console weaponization to learn and understand how to tackle this issue. Research in the ability to analyze consoles under the premise of a digital forensics investigation exists, but little is explored in the actual malicious capabilities of these devices. If these consoles can be weaponized in the experimental setting, then there is the possibility of more skilled cybercriminals attempting to do this in the real world as well. This issue must be addressed before a real incident involving a weaponized video game console occurs. Exploring this possibility can help investigators and cybercrime specialists understand how to counteract these cyberweapons, as well as analyze them to gather evidence for the investigation.

Exploring the possibility of weaponizing video game consoles is only one step towards improving digital forensics research on this topic. One other aspect of video games that needs to be addressed is the social aspect. Many video games now include online interactions, whether it is to form groups to complete tasks or strategizing with other players to face challenges. Communication occurs through text or voice channels that are available in the game. Examples of this can be found in World of Warcraft (WoW), when raid groups are formed, League of Legends, when creating a pre-made lobby, or in Overwatch, once your team has been banded. However, with the massive number of players and the additional network packets being sent out while playing the game, tracking voice communications can be difficult. As a result, there is the possibility of harassment and cyberbullying occurring while playing a game with online interactions. Additionally, video game communities can be formed outside of the game on social platforms, such as Discord or Reddit. While game companies acknowledge the presence of toxicity in online communications, they are unable to do more than have filters and flags for specific buzzwords indicative of negative attitude and hate speech, as well as giving players the ability to report others for such behavior. However, the latter has been reported to be ineffective in preventing toxic behavior online, further emphasizing the rampant toxicity. The limitations of monitoring communication in online video games and social platforms have also provided cybercriminals with opportunities for covert communication.

Sometimes, these online communications can lead to serious real-life events occurring, such as stalking or even murder. Recently, such incidents occurred in Kansas (Grinburg, 2018), Michigan (Harris, 2022), South Korea (Jun-tae, 2021; Naves, 2021), and the United Kingdom (Halliday, 2014; Ngak, 2011). While toxic behavior needs to be addressed, in the context of digital forensics, the challenges presented with online communications mean that even cybercriminals and terrorists can get away with openly communicating with one another in-game. Some online multiplayer games, such as WoW, Minecraft, and Call of Duty, were found to not only have spies and terrorists communicating with each other but were also used as recruitment grounds by extremist groups (Mazzetti and Elliot, 2013; Miller and Silva, 2021; Townsend, 2021). Therefore, digital forensics researchers should also turn their attention to online video game interactions. One potential solution that may help advance research in this area is to create databases about current and past terrorist and extremist phrases and methods and modify existing forensic tools to automatically flag possible malicious communication for specialists to investigate. This can be facilitated with artificial intelligence. Doing so will not only assist in the hunt for malicious communication of any kind but will also improve knowledge management.

A summary of the challenges faced and their potential solutions can be found in Table 4.3.

TABLE 4.3
The Challenges Mentioned and Some Potential Solutions

Challenge	Potential Solution
Some malicious activities are performed through PC games.	• PC games, to an extent, can be treated like standard applications since they leave behind artifacts that may be analyzable using forensic tools.
Consoles may be weaponized for specific malicious activities.	• Invest more research effort into understanding and forensically analyzing video game consoles. • Develop tools that can assist investigators specifically with console analysis.
The massive number of server traffic makes it hard to efficiently monitor for malicious or covert traffic. Malicious online communication may cause or correlate with real-world events.	• Databases can be created from past and currently known terrorist and extremist behavior and methods to modify existing forensic and monitoring tools. This can be facilitated through the integration of artificial intelligence. This will also improve knowledge management and sharing.

5 CONCLUSION

The continuous rise in popularity of video games and the opportunities for growth they provide have drawn much attention from developers and researchers worldwide. However, like any other topic in IT research, cybercriminals and terrorists have found ways to take advantage of these opportunities for their own malicious benefits. From remote script executions and disruptions in service to spying and extremist group recruiting, for motives ranging from monetary gain to simple entertainment, cybercriminals will continue to follow the video game trend as it continues to evolve. Investigators will need to understand how to approach these issues and analyze the devices involved to keep up.

Research work regarding video games in the context of digital forensics so far has mainly focused on preparing guides for analyzing certain aspects of video games, such as custom map features or the forensic acquisition of the game consoles themselves. However, most of the work included in this study focused on aspects of video games related to the standard PC, such as creating education modules or studying traditional cybercrimes.

Future research in digital forensics that focuses on video games should put more consideration on consoles, especially since, at the time of writing this paper, no work has been found on current-generation consoles (PlayStation 5 and Xbox Series X/S). As newer generations of consoles continue to be released, the operating and file systems on these devices continue to become more complex. Investing more time into researching these consoles can help investigators be more prepared when encountering them during an investigation. Additional research should be done on the potential weaponization of video game consoles and related devices, similar to how PC devices (such as USBs or hard disks) and mobile phones can be modified specifically for malicious purposes.

REFERENCES

AbleGamers, n.d. The AbleGamers Charity [WWW Document]. *The AbleGamers Charity*. https://ablegamers.org/

Abrams, L., 2016. World of Warcraft Scam Allows Attackers to Take Control of Victim's User Interface. Bleeping Computer. www.bleepingcomputer.com/news/security/world-of-warcraft-scam-allows-attackers-to-take-control-of-victims-user-interface/

Ahmad, M.A., Keegan, B., Sullivan, S., Williams, D., Srivastava, J., Contractor, N., 2011. Illicit Bits: Detecting and Analyzing Contraband Networks in Massively Multiplayer Online Games, in: 2011 IEEE Third Int'l Conference on Privacy, Security, Risk and Trust and 2011 IEEE Third Int'l Conference on Social Computing. Presented at the 2011 IEEE Third Int'l Conference on Privacy, Security, Risk and Trust (PASSAT) / 2011 IEEE Third Int'l Conference on Social Computing (SocialCom), IEEE, Boston, MA, USA, pp. 127–134. https://doi.org/10.1109/PASSAT/SocialCom.2011.183

Al-Haj, A.M., 2019. *Forensics Analysis of Xbox One Game Console*. arXiv preprint arXiv:1904.00734 10.

American Red Cross, n.d. Stream Hope, Save Lives [WWW Document]. American Red Cross. www.redcross.org/donations/ways-to-donate/play-games-and-fundraise.html

"Archimtiros", 2021. Malicious WeakAura Replaces Auction House Purchases with Overpriced Scams. *Wowhead TBC.* https://tbc.wowhead.com/news/malicious-weakaura-replaces-auction-house-purchases-with-overpriced-scams-322567

Barr-Smith, F., Farrant, T., Leonard-Lagarde, B., Rigby, D., Rigby, S., Sibley-Calder, F., 2021. Dead Man's Switch: Forensic Autopsy of the Nintendo Switch. *Forensic Science International: Digital Investigation* 36, 301110. https://doi.org/10.1016/j.fsidi.2021.301110

bin Mohd Isa, H.A., 2009. *The Practical Analysis Towards Developing a Guideline for the Xbox 360 Forensic.* pdf (Master of Computer Science). Universiti Teknologi Malaysia.

Birnbaum, I., 2016. A Gold-Stealing Script Is Rapidly Spreading Across "World of Warcraft." *Vice.* www.vice.com/en/article/3dapb9/a-virus-is-spreading-across-world-of-warcraft-stealing-gold

Blauw, F.F., Leung, W.S., 2018. ForenCity: A Playground for Self-Motivated Learning in Computer Forensics, in: Drevin, L., Theocharidou, M. (Eds.), *Information Security Education – Towards a Cybersecure Society, IFIP Advances in Information and Communication Technology*. Springer International Publishing, Cham, pp. 15–27. https://doi.org/10.1007/978-3-319-99734-6_2

Blazic, A.J., Cigoj, P., Blazic, B.J., 2016. Serious Game Design for Digital Forensics Training, in: 2016 Third Int'l Conference on Digital Information Processing, Data Mining, and Wireless Communications (DIPDMWC). Presented at the 2016 Third Int'l Conference on Digital Information Processing, Data Mining, and Wireless Communications (DIPDMWC), IEEE, Moscow, Russia, pp. 211–215. https://doi.org/10.1109/DIPDMWC.2016.7529391

Boldi, A., Rapp, A., 2021. Commercial Video Games as a Resource for Mental Health: A Systematic Literature Review. *Behaviour & Information Technology*, pp. 1–37. https://doi.org/10.1080/0144929X.2021.1943524

Burke, P., Craiger, P., 2007a. Forensic Analysis of Xbox Consoles, in: Craiger, P., Shenoi, S. (Eds.), *Advances in Digital Forensics III, IFIP — The International Federation for Information Processing*. Springer New York, New York, NY, pp. 269–280. https://doi.org/10.1007/978-0-387-73742-3_19

Burke, P.K., Craiger, P., 2007b. Xbox Forensics. *Journal of Digital Forensic Practice* 1, pp. 275–282. https://doi.org/10.1080/15567280701417991

Chalk, A., 2021. After Years of Struggling Against DDoS Attacks, Titanfall is being Removed from Sale. *PC Gamer.* www.pcgamer.com/after-years-of-struggling-against-ddos-attacks-titanfall-is-being-removed-from-sale/

Chen, L., Shashidhar, N., Rawat, D., Yang, M., Kadlec, C., 2016. Investigating the Security and Digital Forensics of Video Games and Gaming Systems: A Study of PC Games and PS4 Console, in: 2016 Int'l Conference on Computing, Networking and Communications (ICNC). Presented at the 2016 Int'l Conference on Computing, Networking, and Communications (ICNC), IEEE, Kauai, HI, USA, pp. 1–5. https://doi.org/10.1109/ICCNC.2016.7440557

Cluely, G., 2019. Derp! DDoS Attacker who Brought Down EA, Sony, and Steam Jailed for 27 Months. *Bitdefender.* www.bitdefender.com/blog/hotforsecurity/derp-ddos-attacker-who-brought-down-ea-sony-and-steam-jailed-for-27-months

Comeras-Chueca, C., Marin-Puyalto, J., Matute-Llorente, A., Vicente-Rodriguez, G., Casajus, J.A., Gonzalez-Aguero, A., 2021. The Effects of Active Video Games on Health-Related Physical Fitness and Motor Competence in Children and Adolescents with Healthy Weight: A Systematic Review and Meta-Analysis. *IJERPH* 18, 6965. https://doi.org/10.3390/ijerph18136965

Conrad, S., Dorn, G., Craiger, P., 2010. Forensic Analysis of a PlayStation 3 Console, in: Chow, K.-P., Shenoi, S. (Eds.), *Advances in Digital Forensics VI, IFIP Advances in Information and Communication Technology*. Springer Berlin Heidelberg, Berlin, Heidelberg, pp. 65–76. https://doi.org/10.1007/978-3-642-15506-2_5

Conrad, S., Rodriguez, C., Marberry, C., Craiger, P., 2009. Forensic Analysis of the Sony PlayStation Portable, in: Peterson, G., Shenoi, S. (Eds.), *Advances in Digital Forensics V, IFIP Advances in Information and Communication Technology*. Springer Berlin Heidelberg, Berlin, Heidelberg, pp. 119–129. https://doi.org/10.1007/978-3-642-04155-6_9

Conway, A., James, J.I., Gladyshev, P., 2015. Development and Initial User Evaluation of a Virtual Crime Scene Simulator Including Digital Evidence, in: James, J.I., Breitinger, F. (Eds.), *Digital Forensics and Cyber Crime, Lecture Notes of the Institute for Computer Sciences, Social Informatics and Telecommunications Engineering*. Springer International Publishing, Cham, pp. 16–26. https://doi.org/10.1007/978-3-319-25512-5_2

Cravel, C., 2015. *Xbox One File System Data Storage a Forensic Analysis*. Purdue University.

Crellin, J., Karatzouni, S., 2009. Simulation in Digital Forensic Education. Presented at the Third Int'l Conference on Cybercrime Forensic Education and Training.

Criddle, C., 2021. Cyberpunk 2077 Makers CD Projekt Hit by Ransomware Hack. *BBC News*. www.bbc.com/news/technology-55994787

Davies, M., Read, H., Xynos, K., Sutherland, I., 2015. Forensic Analysis of a Sony PlayStation 4: A First Look. *Digital Investigation* 12, pp. S81–S89. https://doi.org/10.1016/j.diin.2015.01.013

Drachen, A., Sifa, R., Bauckhage, C., Thurau, C., 2012. Guns, Swords and Data: Clustering of Player Behavior in Computer Games in the Wild, in: 2012 IEEE Conference on Computational Intelligence and Games (CIG). Presented at the 2012 IEEE Conference on Computational Intelligence and Games (CIG), IEEE, Granada, Spain, pp. 163–170. https://doi.org/10.1109/CIG.2012.6374152

Drakou, M., Lanitis, A., 2016. On the Development and Evaluation of a Serious Game for Forensic Examination Training, in: 2016 18th Mediterranean Electrotechnical Conference (MELECON). Presented at the 2016 18th Mediterranean Electrotechnical Conference (MELECON), IEEE, Lemesos, Cyprus, pp. 1–6. https://doi.org/10.1109/MELCON.2016.7495415

Ebert, L.C., Nguyen, T.T., Breitbeck, R., Braun, M., Thali, M.J., Ross, S., 2014. The Forensic Holodeck: An Immersive Display for Forensic Crime Scene Reconstructions. *Forensic Sci Med Pathol* 10, pp. 623–626. https://doi.org/10.1007/s12024-014-9605-0

Ebrahimi, M., Lei Chen, 2014. Emerging Cyberworld Attack Vectors: Modification, Customization, Secretive Communications, and Digital Forensics in PC Video Games, in: 2014 Int'l Conference on Computing, Networking and Communications (ICNC). Presented at the 2014 Int'l Conference on Computing, Networking and Communications (ICNC), IEEE, Honolulu, HI, USA, pp. 939–944. https://doi.org/10.1109/ICCNC.2014.6785463

Elfving, V., Lidstrom, R., 2020. *An exploratory forensic analysis of the Xbox One S All Digital*. Halmstad University.

Extra Life, n.d. Extra Life Homepage [WWW Document]. Extra Life: A Program of Children's Miracle Network Hospitals. www.extra-life.org/

Freedman, A., 2022. *Windows on Steam Deck: Benchmarks and Impressions*. Tom's Hardware.

Fuchs, M., 2012. Social Games: Privacy and Security, in: Hercheui, M.D., Whitehouse, D., McIver, W., Phahlamohlaka, J. (Eds.), *ICT Critical Infrastructures and Society, IFIP*

Advances in Information and Communication Technology. Springer Berlin Heidelberg, Berlin, Heidelberg, pp. 330–337. https://doi.org/10.1007/978-3-642-33332-3_30

Games Aid, n.d. Welcome to Games Aid [WWW Document]. *Games Aid: The Games Industry Charity*. www.gamesaid.org/

Gibbs, C., Shashidhar, N., 2015. StegoRogue: Steganography in Two-Dimensional Video Game Maps. *Advances in Computer Science: An International Journal* 4, 6.

Goodin, D., 2014. World of Warcraft Users Hit by Account-Hijacking Malware Attack. *Ars Technica*. https://arstechnica.com/information-technology/2014/01/world-of-warcraft-users-hit-by-account-hijacking-malware-attack/

Gordillo, A., Lopez-Fernandez, D., Tovar, E., 2022. Comparing the Effectiveness of Video-Based Learning and Game-Based Learning Using Teacher-Authored Video Games for Online Software Engineering Education. *IEEE Trans. Educ*. pp. 1–9. https://doi.org/10.1109/TE.2022.3142688

Greig, J., 2022. EA Confirms Dozens of High-Profile FIFA Accounts Hacked. *ZDNet*. www.zdnet.com/article/ea-confirms-dozens-of-high-profile-fifa-accounts-hacked-blame-customer-experience-employees/

Grinburg, E., 2018. Shooting Death in Video Game Leads to a Real One in Kansas. *CNN*. https://edition.cnn.com/2018/01/30/us/kansas-swatting-death-affidavit/index.html

Groupees, n.d. Groupees–Bundles of Joy [WWW Document]. *Groupees*. https://groupees.com/

Grustniy, L., 2021. A Trick Auction to Steal Gold in World of Warcraft. *Kaspersky Daily*. www.kaspersky.com/blog/wow-weakauras-auction-scam/40280/

Hagan, K., 2021. The Explosive Growth of Esports: Will This Sector Take Over? *Medium*. https://medium.com/included-vc/the-explosive-growth-of-esports-will-this-sector-take-over-6a9ee903523c

Hale, C., Chen, L., Liu, Q., 2012. A New Villain: Investigating Steganography in Source Engine Based Video Games, in: Proceedings of the 2012 Hong Kong Int'l Conference on Engineering & Applied Science (HKICEAS). p. 36.

Halliday, J., 2014. *Teenage computer engineer pleads guilty to murdering Breck Bednar, 14*. The Guardian.

Hamilton, I.A., 2020. Gaming Giant Capcom Says the Data of up to 350,000 People, Including Players, was Stolen in a Massive Ransomware Attack. *Business Insider*. www.businessinsider.com/capcom-hack-ransomware-attack-hackers-customer-data-2020-11

Harris, C., 2022. Michigan Man Allegedly Strangles Boyfriend of 10 Years During Argument over Video Game. *People*. https://people.com/crime/michigan-man-allegedly-strangles-longtime-boyfriend-during-argument-over-video-game/

Help for Heroes, n.d. How to do a Charity Gaming Stream [WWW Document]. *Help for Heroes*. www.helpforheroes.org.uk/give-support/fundraise/hero-up/how-to-do-a-charity-gaming-stream/

Herr, C., Allen, D., 2015. Video Games as a Training Tool to Prepare the Next Generation of Cyber Warriors, in: Proceedings of the 2015 ACM SIGMIS Conference on Computers and People Research. Presented at the SIGMIS-CPR '15: 2015 Computers and People Research Conference, ACM, Newport Beach California USA, pp. 23–29. https://doi.org/10.1145/2751957.2751958

Hollister, S., 2020. The FDA Just Approved the First Prescription Video Game — It's for Kids with ADHD. *The Verge*. www.theverge.com/2020/6/15/21292267/fda-adhd-video-game-prescription-endeavor-rx-akl-t01-project-evo

Humble Bundle, n.d. Humble Charities [WWW Document]. *Humble Bundle*. www.humblebundle.com/charities?hmb_source=navbar

Irmler, F., Creutzburg, R., 2014. Possibilities for Retracing of Copyright Violations on Current Video Game Consoles by Optical Disk Analysis, Presented at the IS&T/SPIE Electronic

Imaging, San Francisco, California, USA, p. 90300G. https://doi.org/10.1117/12.2044933

Jin, G., Tu, M., Kim, T.-H., Heffron, J., White, J., 2018. Evaluation of Game-Based Learning in Cybersecurity Education for High School Students. *EduLearn* 12, pp. 150–158. https://doi.org/10.11591/edulearn.v12i1.7736

Jun-tae, K., 2021. *Stalker, murderer of three sentenced to life in prison*. The Korea Herald.

Karabiyik, U., Mousas, C., Sirota, D., Iwai, T., Akdere, M., 2019. A Virtual Reality Framework for Training Incident First Responders and Digital Forensic Investigators, in: Bebis, G., Boyle, R., Parvin, B., Koracin, D., Ushizima, D., Chai, S., Sueda, S., Lin, X., Lu, A., Thalmann, D., Wang, C., Xu, P. (Eds.), *Advances in Visual Computing, Lecture Notes in Computer Science*. Springer International Publishing, Cham, pp. 469–480. https://doi.org/10.1007/978-3-030-33723-0_38

Karmali, L., 2014. Blizzard Warns of Trojan Bypassing Warcraft Authenticators. *IGN*. www.ign.com/articles/2014/01/06/blizzard-warns-of-trojan-bypassing-warcraft-authenticators

Kelly, C., Schuler, S., Johnson, P., 2021. Gaming: The Next Super Platform. *Accenture*. www.accenture.com/us-en/insights/software-platforms/gaming-the-next-super-platform?c=acn_glb_thenewgamingexpbusinesswire_12160747%26n=mrl_0421

Khanji, S., Jabir, R., Iqbal, F., Marrington, A., 2016. Forensic Analysis of Xbox One and PlayStation 4 Gaming Consoles, in: 2016 IEEE Int'l Workshop on Information Forensics and Security (WIFS). Presented at the 2016 IEEE Int'l Workshop on Information Forensics and Security (WIFS), IEEE, Abu Dhabi, United Arab Emirates, pp. 1–6. https://doi.org/10.1109/WIFS.2016.7823917

Kim, M., 2021. New World Reportedly Has a Vulnerability That Makes It Possible To Crash Players Through the Text Box. *IGN*. www.ign.com/articles/new-world-text-box-html-vulnerability-game-crash

Kowal, M., Conroy, E., Ramsbottom, N., Smithies, T., Toth, A., Campbell, M., 2021. Gaming Your Mental Health: A Narrative Review on Mitigating Symptoms of Depression and Anxiety Using Commercial Video Games. *JMIR Serious Games* 9, e26575. https://doi.org/10.2196/26575

Kwon, H., Mohaisen, A., Woo, J., Kim, H.K., Kim, Y., Lee, E.J., 2016. Crime Scene Reconstruction: Online Gold Farming Network Analysis. *IEEE Trans.Inform.Forensic Secur*. 1–1. https://doi.org/10.1109/TIFS.2016.2623586

Litchfield, T., 2022. Dark Souls PvP Servers are Down as Security Vulnerability is Investigated. *PC Gamer*. www.pcgamer.com/psa-dont-play-dark-souls-3-until-a-new-remote-code-execution-vulnerability-is-patched/

Lopez-Fernandez, D., Gordillo, A., Alarcon, P.P., Tovar, E., 2021. Comparing Traditional Teaching and Game-Based Learning Using Teacher-Authored Games on Computer Science Education. *IEEE Trans. Educ*. 64, pp. 367–373. https://doi.org/10.1109/TE.2021.3057849

Luttenberger, S., Kröger, K., Creutzburg, R., 2012. Remarks on Forensically Interesting Microsoft XBox 360 Console Features, Presented at the IS&T/SPIE Electronic Imaging, Burlingame, California, USA, p. 83040Q. https://doi.org/10.1117/12.909675

Martinez, L., Gimenes, M., Lambert, E., 2022. Entertainment Video Games for Academic Learning: A Systematic Review. *Journal of Educational Computing Research* 073563312110538. https://doi.org/10.1177/07356331211053848

May, E., 2021. Streamlabs and Stream Hatchet Q3 2021 Live Streaming Industry Report. *Streamlabs*. https://streamlabs.com/content-hub/post/streamlabs-and-stream-hatchet-q3-2021-live-streaming-industry-report

Mazzetti, M., Elliot, J., 2013. Spies Infiltrate a Fantasy Realm of Online Games. *The New York Times*. www.nytimes.com/2013/12/10/world/spies-dragnet-reaches-a-playing-field-of-elves-and-trolls.html

Miller, C., Silva, S., 2021. Extremists Using Video-Game Chats to Spread Hate. *BBC News*. www.bbc.com/news/technology-58600181

Moore, J., Baggili, I., Marrington, A., Rodrigues, A., 2014. Preliminary Forensic Analysis of the Xbox One. *Digital Investigation* 11, pp. S57–S65. https://doi.org/10.1016/j.diin.2014.05.014

Morris, S., 2012. Virtual Crime: Forensic Artefacts from Second Life, in: Cybercrime Forensics Education and Training.

Munir, S., Baig, M.S.I., 2019. *Challenges and Security Aspects of Blockchain Based Online Multiplayer Games*. Halmstad University.

Naves, A.W., 2021. Online Gamer and Family Killed by Spurned Stalker. *Medium*.

Ngak, C., 2011. Man Chokes Kid for "Beating" Him in Call of Duty Game. *CBS News*.

Nordhaug, Ø.A., 2014. *The Forensic Challenger–A Digital Forensic E-Learning Platform*. Gjovik University College.

Oliveira, C.B., Pinto, R.Z., Saraiva, B.T.C., Tebar, W.R., Delfino, L.D., Franco, M.R., Silva, C.C.M., Christofaro, D.G.D., 2020. Effects of Active Video Games on Children and Adolescents: A Systematic Review with Meta-analysis. *Scand J Med Sci Sports* 30, pp. 4–12. https://doi.org/10.1111/sms.13539

Pan, Y., Mishra, S., Yuan, B., Stackpole, B., Schwartz, D., 2012. Game-Based Forensics Course for First Year Students, in: Proceedings of the 13th Annual Conference on Information Technology Education–SIGITE '12. Presented at the 13th annual conference, ACM Press, Calgary, Alberta, Canada, p. 13. https://doi.org/10.1145/2380552.2380558

Pan, Y., Schwartz, D., Mishra, S., 2015. Gamified Digital Forensics Course Modules for Undergraduates, in: 2015 IEEE Integrated STEM Education Conference. Presented at the 2015 IEEE Integrated STEM Education Conference (ISEC), IEEE, Princeton, NJ, USA, pp. 100–105. https://doi.org/10.1109/ISECon.2015.7119899

Peñuelas-Calvo, I., Jiang-Lin, L.K., Girela-Serrano, B., Delgado-Gomez, D., Navarro-Jimenez, R., Baca-Garcia, E., Porras-Segovia, A., 2022. Video Games for the Assessment and Treatment of Attention-Deficit/Hyperactivity Disorder: A Systematic Review. *Eur Child Adolesc Psychiatry* 31, pp. 5–20. https://doi.org/10.1007/s00787-020-01557-w

Pessolano, G., Read, H.O.L., Sutherland, I., Xynos, K., 2019. Forensic Analysis of the Nintendo 3DS NAND. *Digital Investigation* 29, pp. S61–S70. https://doi.org/10.1016/j.diin.2019.04.015

Podhradsky, D.A.L., D'Ovidio, R., Casey, C., 2011a. Identity Theft and Used Gaming Consoles: Recovering Personal Information from Xbox 360 Hard Drives. Presented at the AMCIS 2011, p. 18.

Podhradsky, A.L., D'Ovidio, R., Casey, C., 2011b. A Practitioners Guide to the Forensic Investigation of Xbox 360 Gaming Consoles. Presented at the Annual ADFSL Conference on Digital Forensics, Security and Law, p. 19.

Podhradsky, A., D'Ovidio, R., Casey, C., 2012. The XBOX 360 and Steganography: How Criminals and Terrorists Could Be "Going Dark." Presented at the Annual ADFSL Conference on Digital Forensics, Security and Law.

Podhradsky, A., D'Ovidio, R., Engebretson, P., Casey, C., 2013. Xbox 360 Hoaxes, Social Engineering, and Gamertag Exploits, in: 2013 46th Hawaii Int'l Conference on System Sciences. Presented at the 2013 46th Hawaii Int'l Conference on System Sciences (HICSS), IEEE, Wailea, HI, USA, pp. 3239–3250. https://doi.org/10.1109/HICSS.2013.633

Purslow, M., 2022. Multiple High-Profile FIFA Ultimate Team Traders Hacked in the Same Week. *IGN*. www.ign.com/articles/fifa-ultimate-team-trader-hack

Rakitianskaia, A.S., Olivier, M.S., Cooper, A.K., 2011. Nature and Forensic Investigation of Crime in Second Life. Presented at the 10th Annual Information Security South Africa Conference, p. 9.

Ray, S., 2021. Cyberpunk 2077 Maker CD Projekt Red Has Been Hacked, Says It Won't Pay Ransom. *Forbes*. www.forbes.com/sites/siladityaray/2021/02/09/cyberpunk-2077-maker-cd-projekt-red-has-been-hacked-says-it-wont-pay-ransom/?sh=6bd3336a4896

Read, H., Thomas, E., Sutherland, I., Xynos, K., Burgess, M., 2016. A Forensic Methodology for Analyzing Nintendo 3DS Devices, in: Peterson, G., Shenoi, S. (Eds.), *Advances in Digital Forensics XII, IFIP Advances in Information and Communication Technology*. Springer International Publishing, Cham, pp. 127–143. https://doi.org/10.1007/978-3-319-46279-0_7

Robertson, A., 2018. The Man Behind a Spree of Gaming Network Cyberattacks has Pleaded Guilty. *The Verge*. www.theverge.com/2018/11/7/18071764/austin-thompson-derptroll ing-sony-blizzard-game-ddos-arrest-guilty

Roth, E., 2022. Dark Souls 3 Exploit Could Let Hackers Take Control of Your Entire Computer. *The Verge*. www.theverge.com/2022/1/22/22896785/dark-souls-3-remote-execution-exploit-rce-exploit-online-hack

Santos, I.K. dos, Medeiros, R.C. da S.C. de, Medeiros, J.A. de, Almeida-Neto, P.F. de, Sena, D.C.S. de, Cobucci, R.N., Oliveira, R.S., Cabral, B.G. de A.T., Dantas, P.M.S., 2021. Active Video Games for Improving Mental Health and Physical Fitness—An Alternative for Children and Adolescents during Social Isolation: An Overview. *IJERPH* 18, p. 1641. https://doi.org/10.3390/ijerph18041641

Schofield, D., 2011. Playing with Evidence: Using Video Games in the Courtroom. *Entertainment Computing* 2, pp. 47–58. https://doi.org/10.1016/j.entcom.2011.03.010

Schofield, D., 2009. Animating Evidence Computer Game Technology in the Courtroom. *Journal of Information, Law, and Technology* 1, pp. 1–21.

Snider, M., 2021. Two-Thirds of Americans, 227 million, Play Video Games. For Many Games were an Escape, Stress Relief in Pandemic. *USA Today*. www.usatoday.com/story/tech/gaming/2021/07/13/video-games-2021-covid-19-pandemic/7938713002/

St. Jude's Children's Research Hospital, n.d. St. Jude PLAY LIVE [WWW Document]. *St. Jude Play Live*. www.stjude.org/get-involved/other-ways/video-game-charity-event.html

Tabuyo-Benito, R., Bahsi, H., Peris-Lopez, P., 2019. Forensics Analysis of an On-line Game over Steam Platform, in: Breitinger, F., Baggili, I. (Eds.), *Digital Forensics and Cyber Crime, Lecture Notes of the Institute for Computer Sciences, Social Informatics and Telecommunications Engineering*. Springer International Publishing, Cham, pp. 106–127. https://doi.org/10.1007/978-3-030-05487-8_6

Tassi, P., 2015. How ISIS Terrorists May Have Used PlayStation 4 to Discuss and Plan Attacks. *Forbes*. www.forbes.com/sites/insertcoin/2015/11/14/why-the-paris-isis-terrorists-used-ps4-to-plan-attacks/?sh=6ea0d4407055

Taylor, D.C.P.J., Mwiki, H., Dehghantanha, A., Akibini, A., Choo, K.K.R., Hammoudeh, M., Parizi, R., 2019. Forensic Investigation of Cross Platform Massively Multiplayer Online Games: Minecraft as a Case Study. *Science & Justice* 59, pp. 337–348. https://doi.org/10.1016/j.scijus.2019.01.005

Tidy, J., 2020. Capcom Hack: Up to 350,000 People's Information Stolen. *BBC News*. www.bbc.com/news/technology-54958782

Townsend, M., 2021. How Far Right Uses Video Games and Tech to Lure and Radicalise Teenage Recruits. *The Guardian*. www.theguardian.com/world/2021/feb/14/how-far-right-uses-video-games-tech-lure-radicalise-teenage-recruits-white-supremacists

Turnbull, D.B., 2008. *Forensic Investigation of the Nintendo Wii: A First Glance 2*, p. 8.

Twitch, n.d. Charity Fundraising on Twitch [WWW Document]. *Twitch Creator Camp*. www.twitch.tv/creatorcamp/en/connect-and-engage/charity-streaming/

van Voorst, R., Kechadi, M.-T., Le-Khac, N.-A., 2015. Forensic Acquisition of IMVU: A Case Study. *JDFSL* 10. https://doi.org/10.15394/jdfsl.2015.1212

Vaughan, C., 2004. Xbox Security Issues and Forensic Recovery Methodology (utilising Linux). *Digital Investigation* 1, pp. 165–172. https://doi.org/10.1016/j.diin.2004.07.006

WePC, 2022. Video Game Industry Statistics, Trends and Data In 2022, *WePC*. www.wepc.com/news/video-game-statistics/

West, M., 2015. *Confronting Criminal Enterprise in Online Gaming: Determining Awareness and Capabilities Regarding the Emergence of Online Gaming Cybercrime*. Utica College.

Wijaya, E., Khanifa, F., Wijanarka, H., Maulana, S.I., 2014. Approach Method to Follow Up Fraud in Online Gaming, in: 2014 Int'l Conference on Information Technology Systems and Innovation (ICITSI). Presented at the 2014 Int'l Conference on Information Technology Systems and Innovation (ICITSI), IEEE, Bandung, Indonesia, pp. 153–158. https://doi.org/10.1109/ICITSI.2014.7048256

Wijman, T., 2021. The Games Market and Beyond in 2021: The Year in Numbers. *Newzoo*. https://newzoo.com/insights/articles/the-games-market-in-2021-the-year-in-numbers-esports-cloud-gaming/

Wijman, T., 2020. Three Billion Players by 2023: Engagement and Revenues Continue to Thrive Across the Global Games Market. *Newzoo*. https://newzoo.com/insights/articles/games-market-engagement-revenues-trends-2020-2023-gaming-report/

Williams, L., 2022. A Pandemic Is a Dream Come True for Gamers. *Bloomberg Opinion*. www.bloomberg.com/opinion/articles/2022-01-16/pandemic-s-boost-for-video-game-industry-is-a-dream-come-true-kyh9nekz

Winkie, L., 2021. Titanfall 2 was Abandoned by EA, and Then Things Got Weird. *IGN*. www.ign.com/articles/titanfall-2-hack-saga-feature

Xynos, K., Harries, S., Sutherland, I., Davies, G., Blyth, A., 2010. Xbox 360: A Digital Forensic Investigation of the Hard Disk Drive. *Digital Investigation* 6, 104–111. https://doi.org/10.1016/j.diin.2010.02.004

Yarramreddy, A., Gromkowski, P., Baggili, I., 2018. Forensic Analysis of Immersive Virtual Reality Social Applications: A Primary Account, in: 2018 IEEE Security and Privacy Workshops (SPW). Presented at the 2018 IEEE Security and Privacy Workshops (SPW), IEEE, San Francisco, CA, pp. 186–196. https://doi.org/10.1109/SPW.2018.00034

Yerby, J., Hollifield, S., Kwak, M., Floyd, K., 2014. Development of Serious Games for Teaching Digital Forensics. *IIS*. https://doi.org/10.48009/2_iis_2014_335-343

5 Dances with the Illuminati

Hands-On Social Engineering in Classroom Setting

Sten Mäses, Birgy Lorenz, Kaido Kikkas and Kristjan Karmo
Tallinn Institute of Technology
Tallinn, Estonia

1 INTRODUCTION

Online phishing scams are annoying, dangerous, and almost impossible to prevent (Sahingoz et al, 2019). However, they present a unique opportunity for educational exploitation. In this chapter, we suggest that if controlled and properly supervised, replying to phishing attempts can be a way to acquire hands-on experience about

DOI: 10.1201/9781003415060-7

the criminal *modus operandi*. Furthermore, having contact with real scammers can provide an engaging way to introduce various cybersecurity-related topics to students.

2 RELATED WORK

One of the most important teaching challenges in the field of cybersecurity is to create the best learning simulations that meet real-life standards. One needs to learn the rules (i.e., related legislation), but also test skills against real perpetrators. It is also important to learn about the old ways and their modern counterparts that go against the human core principles of helping others and being part of a group (as exemplified by the ancient ninjutsu parallels drawn by Wilhelm and Andress, 2011). However, testing learning simulations in real life is difficult due to ethical issues. At the same time, online criminals need care about neither laws nor ethics. Students of cybersecurity (at least on the Master's level) need to have real-life experience to be ready for their future job roles. Considering this, we have built upon the academic inquiry tradition and directed our study tasks in a gray area, which is still academically valid (in the legal and ethical sense). We have been inspired by psychologists like Cialdini (2014) and his principles of influence (especially the "click-whirr" automatic responses). Additionally, other treatises on social engineering should be credited, such as Long (2008) and Hadnagy (2011). We have also investigated chain value issues to develop our exercises (Levchenko et al., 2011).

3 THE REAL-LIFE INSPIRATION

It all started with personal curiosity and a conversation with some of the online scammers. One of the authors received a suspicious friend request on a social networking site. What originally looked like a new account of an old relative, turned out to be a scam attempt via identity theft (the fake account was removed shortly after reporting) and the economic uncertainty of the first COVID lockdowns. In a classic pay-to-get-paid scheme the attackers were offering a substantial financial assistance package ($200,000) in exchange for a small fee ($4,500) and some personal data. The mobile number used in the form started receiving SMS loan offers within a few days. After convincing the scammers to give out several PayPal accounts, one of the authors ended the conversation and reported all the involved compromised accounts to the relevant authorities. This was the first half of the inspiration.

The second story started about a month later with another friend request. This time, the message offered one of the authors of this research untold riches and power by joining the Illuminati. This time the promised payout was bigger, including $500,000 in cash; a car worth $300,000; a house; $1,000,000 bank transfer every month; appointments with top five world leaders and top five celebrities. The "sacrifice" upon joining was either $10,000 to be paid in two instalments or the life of a living relative. This time the perpetrators were going all-in on greed, not even playing the necessity angle as in the previous encounter. After spending some time acquiring more compromised accounts to report, we ended the conversation. A few hours later, the scammer restarted the conversation, seemingly getting their chat windows mixed up. Then the author decided to state that it has been a scam, and as the scammer used some personal data from the previous encounter, most likely to appear more menacing. As the "personal data" was from a fictitious persona, it became clear that these scams are somehow related, if not conducted by the same scammer. This short exchange then led to the scammer seemingly dropping the façade and switching to yet another tactic, asking for charity instead of playing to greed or threats. The author then attempted to earnestly educate the scammer toward offering their services on Fiverr or similar platforms and was met with another change in tactics, asking to create accounts on different platforms.

These encounters led us striving to better understand the processes and scripts used by scammers, as well as finding more effective ways to divert the scammers' attention away from susceptible victims and educate potential targets about scammers' tactics to make them less vulnerable.

4 MOTIVATION, SCHEDULE, AND FOCUS GROUP

Based on the Illuminati story described above, we developed a set of experimental hands-on homework tasks that used real-life phishing scams to teach human aspects of cybersecurity in a Master's-level course that took place at Tallinn University of Technology in the Fall term of 2021. For the legal and ethical framework, we relied on previous, similar endeavors (see related work section). Aiming to turn those experiences into engaging homework tasks led to a set of assignments: creating a fake identity, gathering and sorting fraudulent email, automatic separation of scam emails, and promoting security awareness via additional advanced steps involving chatting with scammers and preparing an automatic reply to fraud attempts.

The chosen focus group consisted of 57 first-year Master's students of cybersecurity—some of them had an earlier background in IT and cybersecurity, other disciplines included law, management, and psychology. Eighty percent of the students had worked in the field of IT or cybersecurity, and 40% were working during the course (beside their studies). The exercises had to be completed in two months (every exercise had a one-to-three-week deadline). Every student could also decide what part of the exercises they wanted to take part in, and if necessary, negotiate a custom set of tasks based on individual limitations. For instance, chatting with actual scammers was strictly voluntary. Also, there was a possibility of not being part of the series of exercises at all.

5 RESULTS

The homework was chosen by 56 Master's students out of 57 (one decided to leave the course due to personal reasons). The student feedback was mostly positive. While mentioning that the exercise was interesting and fun, several students suggested that more time could be given to complete the tasks, especially for gathering scam messages at the beginning.

TABLE 5.1
Analysis of the Exercises

Exercise	Participants	Average Score
Creating a fake identity	53	79%
Gather and sort fraudulent email	56	64%
Automatic filtering of scam e-emails	48	87.5%
Chatting with the scammer	28	85%
Automatic reply to fraud attempt	12	67%
Raise security awareness, and advanced level exercises	30	66%

The tasks based on actual online scams provided an interesting basis for further discussions. The communication with scammers enabled students to have deeper insights into several important topics of cybersecurity (see Table 5.1).

Creating a fake identity. As communication with scammers might bring unwanted consequences, it makes sense to not use one's real identity. The first task for the students was to create a fictional identity including a computer-generated photo, an email address, and a physical address that does not exist. This brought up several interesting questions about the legality of such a task. As the students confirmed through their research, stealing someone's identity is a punishable, illegal act in most countries.

Creating a previously non-existing fictional identity is usually permitted by the law (Lorenz and Kikkas, 2020)—especially if there will likely be no financial harm done. Also, the students brought out interesting technical aspects for preserving online anonymity (separation from the real identity) during the experiment. Some examples include using a Tor relay for Internet traffic and a virtual machine to minimize any potential harm to the student's equipment. Figure 5.1 illustrates an example of a fictional identity. A combination of https://generated.photos, www.picturando.com/fake/passports, and additional photo editing has been used to generate the picture.

Gathering and sorting scam email. The next step was to use the created email address to gather scam emails. That included posting the email address to suspicious forums. For automatic filtering of scam emails, we used similar techniques to spam filtering.

Chatting with scammers. A previously generated fictional identity can be used to reply to scams. Going further with the conversation enables us to study the process that scammers are using. Criminal journey mapping (as suggested by Somer et al., 2016) can be done based on this. Figure 5.2 displays one of the chats of this kind.

Automatic replying to scams. For the solution to be scalable, the replies to the scam email could be made automatic. The students can create algorithms to generate meaningful patterns for replies.

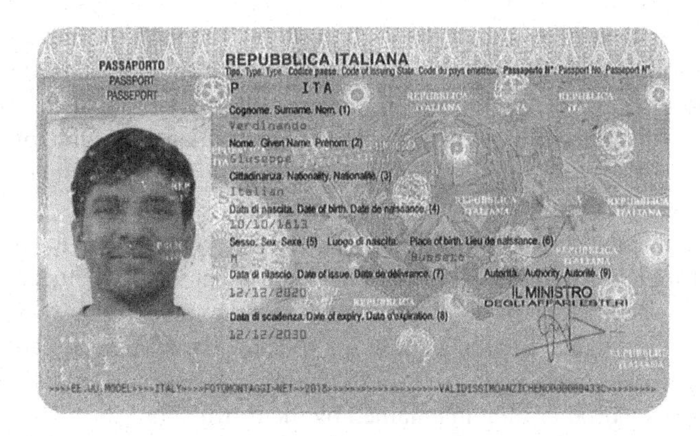

FIGURE 5.1 An example of a fictional identity with a computer-generated photo.

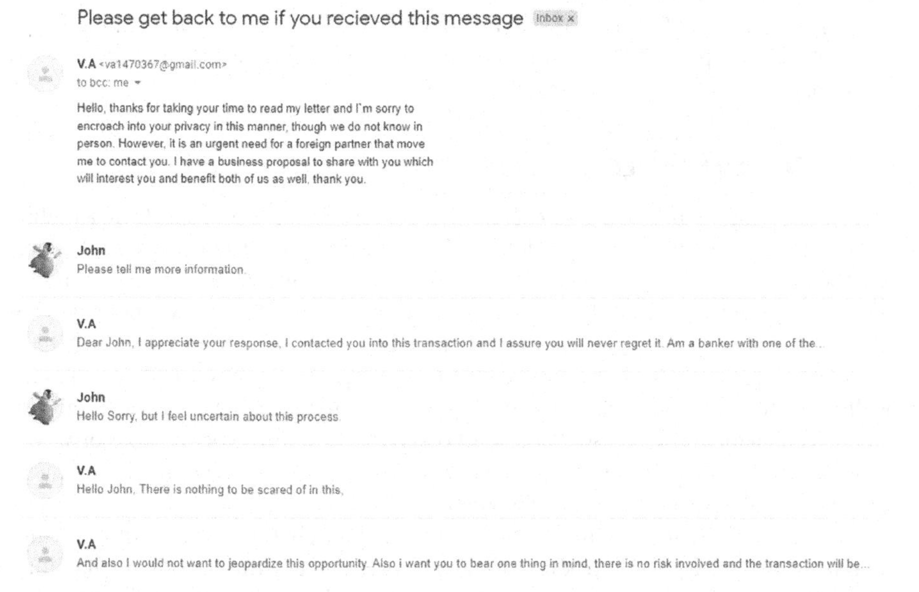

FIGURE 5.2 Excerpt from a conversation of John (fake identity) with a scammer (named V.A).

Raising security awareness. This part of the exercise encouraged students to create materials for raising security awareness of the wider public, including blog articles, social media posts, and videos.

The students produced many innovative solutions:

- Several automatic filtering algorithms were developed in Python, Java, JavaScript, and VBA. The most used machine learning libraries were scikit-learn, TensorFlow, and bag-of-words. The Naive Bayes algorithm was often used for classification.

- Surprisingly, we found that the reply patterns needed to achieve an extended conversation with a scammer did not need to be very sophisticated. Standard replies such as "Please tell me more." or "I did not understand." managed to engage scammers for quite a while. Some challenges highlighted different understanding of terminology, but also reflected differences in required freedom and learning curve.
- The technical tasks did not require any specific technology. In the future, a template for technical tasks could be considered. That way, the students can focus more on specific details and less on choosing a suitable framework.
- Some cybersecurity students chose not to share the awareness campaign posts via their social media accounts. In the future, the best posts could be collected to a course portal or blog maintained by the instructors.
- Several task descriptions were initially unclear for students or offered alternative ways for understanding. It was especially useful to have a forum for specifying requirements and grading logic.
- The concepts of spam and scam were confusing for many. Although the logic of several spam filters can often be adapted to classify scam emails as well, it might not always be reasonable.

6 LESSONS LEARNED

The strong point of these kinds of exercises is that they involve concepts from a multitude of aspects in cybersecurity (including ethical, legal, and technological) that are beneficial for future security experts. Having a real communication with a scammer motivates students to seriously analyze the measures to protect their own privacy. The task of creating a fake identity touches upon several legal and ethical topics (whether it is legal and/or morally justified to create a fake identity and in what conditions) as well as technological solutions. For example, seeing the machine-generated realistic photos can enable further discussions on how to determine whether a photo or video is real or fake, how to ensure a proper chain of evidence when dealing with forensic evidence, and how to protect a digital identity.

The task of creating a machine learning approach for distinguishing scam emails serves as a good introduction to the nuances and limitations of machine learning. For example, in our course, some simple keyword-based filters proved to be more effective in filtering scam emails than complex machine learning algorithms.

The discussion can also include ethics—for instance, what is ethical to teach (e.g., creating fake accounts), especially if this means violating user agreements or other regulations. Psychological influencing/motivation factors can be analyzed based on the manipulation techniques used by the scammers.

Therefore, the hands-on social engineering exercises can serve as a useful tool for making the tasks more interesting and engaging for the students. However, it should be noted that sufficient time must be dedicated to explaining the related ethical aspects. For example, using a fake identity in an exercise should ignite several discussions about the legal consequences of doing so in the wrong context.

7 CONCLUSIONS AND FUTURE WORK

The tasks based on actual online scams provided an interesting basis for further discussions. The communication with scammers enabled students to have deeper insights into several important topics of cybersecurity. Although the tasks connected to real scammers were perceived to have a positive impact on student engagement, further research is required to quantify the more exact impact and determine more specific strategies for using such tasks in various curricula.

REFERENCES

Cialdini, R. B. *Influence: Science and Practice.* 5th Edition. Pearson 2014. ISBN: 978-1-292-02229-1

Hadnagy, C. *Social Engineering: The Art of Human Hacking.* Wiley 2011. ISBN: 978-0-470-63953-5

Levchenko, K., Pitsillidis, A., Chachra, N., Enright, B., Félegyházi, M., Grier, C., ... & Savage, S. 2011, May. Click Trajectories: End-to-end Analysis of the Spam Value Chain. In *2011 IEEE Symposium on Security and Privacy* (pp. 431-446). IEEE.

Long, J. 2008. No Tech Hacking: A Guide to Social Engineering, Dumpster Diving, and Shoulder Surfing. *Syngress.* ISBN 13: 978-1-59749-215-7

Lorenz, B., & Kikkas, K. 2020, July. Pedagogical Challenges and Ethical Considerations in Developing Critical Thinking in Cybersecurity. In *2020 IEEE 20th International Conference on Advanced Learning Technologies* (ICALT) (pp. 262–263). IEEE.

Sahingoz, O. K., Buber, E., Demir, O., & Diri, B. 2019. Machine Learning Based Phishing Detection from URLs. *Expert Systems with Applications*, 117, (pp. 345–357).

Somer, T., Hallaq, B., & Watson, T. 2016, July. Utilising Journey Mapping and Crime Scripting to Combat Cybercrime. In *European Conference on Cyber Warfare and Security* (p. 276). Academic Conferences International Limited.

Wilhelm, T. & Andress, J. 2011. Ninja Hacking: Unconventional Penetration Testing Tactics and Techniques. *Syngress.* ISBN: 978-1-59749-588-2

6 Studying Fake News Proliferation by Detecting Coordinated Inauthentic Behavior

William Emmanuel S. Yu
Ateneo de Manila University
Manila, Philippines

INTRODUCTION

It takes a village to raise a child. This is a famous African proverb used commonly to describe communal efforts to achieve societal goals. The principle is that an entire community takes responsibility for raising a child and not just their immediate parents.

In the paper by Giglietto et al. (2020), the authors describe a number of innovative techniques used to create networks within social media platforms that are used to propagate misinformation and disinformation to sow discord [1]. There have since been many studies looking closely at this new instant and readily accessible platform for spreading information of any kind. This is the new reality we live in. The Internet has allowed unprecedented access to information. But it has also provided a platform for propagating disinformation and misinformation. The Internet is a widely available, wonderful, and open resource. With any shared resource, we do not want to fall prey to British Economist William Foster Lloyd's "*Tragedy of the Commons*". However, it is not just the job of content providers, service providers, and social

DOI: 10.1201/9781003415060-8

media network providers to keep things in check. In this chapter, a safe and secure information ecosystem powered by the Internet also takes a village to keep open, healthy, and alive.

It is technology that allows anybody to become a publisher. True freedom of expression. It is also technology that is used to manipulate this system. At times, in unintended and nefarious ways. In this chapter, we take a step back to look at how information can be manipulated in these technology platforms and how we can detect such interference.

THE PROBLEM OF RELEVANCE

Let us take a step back. The problem of social media is the problem of relevance. If anybody can put information into the platform then how do we ensure that (1) only relevant information gets to its intended audience, and (2) we do not drown them in too much information. At the current world population of over seven billion people with easily five billion people connected to the Internet [2], this is potentially a lot of information. According to the famous Internet minute by Lori Lewis [3], this is easily 197 million emails sent, 414 thousand apps downloaded, 500 hours of content uploaded, and 69 million messages sent. It is quite easy to see in this ocean of content how one can get overwhelmed.

In the early days of mass communications, the printing press allowed us to put compressed and dense amounts of information for widespread distribution in paper. This came in the form of media such as newspapers and books. These types of media were quite convenient as they can contain a dense amount of information that can be consumed at the convenience of the reader. The printing press also allows for mass printing and thus mass distribution of this material. The amount of information was constrained by the limits of physics. For any publication, there is only a limited amount of paper and ink that can be used to ensure wide distribution at a reasonable cost. The solution to this dilemma is curation. Depending on the publication, experts and editors are employed to curate material that enter their publications. The higher the standard of curation with a reputable editorial board results in a higher perception of that publication's quality. Social media changed all that.

Enter the problem of relevance. Instead of editorial boards controlling information that is served in these social media platforms, we now have complex algorithms that aim to serve the role of finding relevant content for the appropriate user. This creates a situation where anybody on these platforms can spread information on these platforms and hope to have them be received by receptive audiences. It is now the algorithms' job to ensure proper targeting.

THE WISDOM OF THE CROWD

Google, a major Internet player, has made headway and cracked the code of relevance in early 1999 with Pagerank [4]. Google is a search engine. Their problem is how to give the most relevant results to a search query. In its simplest form, Pagerank attempts to provide a relevance score to the millions of websites it indexes. To solve this problem, Pagerank provides a score that is determined by the number of pages that link to that particular page given particular search terms (and many other factors). In short, the more pages link to your page, the more relevant the page, the better the score. Surely the more people refer to it the more relevant it is. This does remind me about using citations as the basis of the impact of an academic publication, or the product recommendations in storefronts that are based on the popularity of the product.

The problem of social media platforms is similar. You can think of the information feed served by platforms such as Facebook, Twitter, YouTube, Instagram, Quora, TikTok, and others as results of a search query. Every time you visit these platforms, you make a search. So content on these platforms also needs to be scored. The higher the score the more likely it is to emerge on top of your feed. Relevance is based on popularity. The platforms compete with each other and zealously guard and enhance their algorithms to ensure better relevance. We defer to the wisdom of the crowds enforced by these algorithms.

GAMING THE SYSTEM

Just like the Google Pagerank algorithm, creative people have studied, worked, and endeavored to ensure that their websites score highly on particular searches. This led to the emergence of a multi-million-dollar industry called search engine optimization [5] or SEO. SEO is an attempt to use one's knowledge of the algorithms and the ecosystem it resides in to drive results towards your own goals. In this case, drive results to ensure your website emerges when your desired keywords are searched.

With anything driven by algorithms, these can also be studied. Like SEO, people can study these algorithms to perform actions that exploit their knowledge of how they work. There is a constant battle between people doing SEOs and platforms. However, not all actions are aligned with the rules of the platforms. This is also true on social media platforms. There are parties who aim to game these systems.

In the case of social media platforms, the goal is to ensure that your desired content reaches the audience that it is intended for. This is to push information. There are platform authorized ways to do this such as promote your content (i.e. boosting), get more followers (i.e. please like and subscribe), get more interactions, and many others. These are genuine community building activities within these platforms. However, in the case of misinformation and disinformation, these mechanisms can also be used in not so organic ways. This includes hiring of troll farms (i.e. paid or unpaid), application of bots, or both. This is coordinated inauthentic behavior (CIB).

COORDINATING INAUTHENTIC BEHAVIOR

The study of CIB has been a topic of interest in many information security research circles particularly with respect to tracking down misinformation and disinformation campaigns. The study of CIB is not new. Many platforms such as Facebook have made attempts to clamp down on these practices [6]. The use of these pages or accounts has been instrumental in the spread of disinformation. To limit this type of bad behavior, the core control social media platform uses the "one person, one account" principle. Nothing can get more democratic than this: one person one vote. We can think of algorithms working on constant elections where each of us gets a vote. CIB is essentially creating additional voters with their methods per person. So one person can get many votes.

This is done using many ways. The three most common ways are the use of troll farms, sock puppets, and bots.

Troll Farms are groups of people hired to perform activities that promote the content and material of the farm's employer or benefactor. These troll farm operators or analysts are also called trolls. In the early days of the Internet, one of the biggest groups of networked users was those playing massively multiplayer online role-playing games (MMORPG). One of the most popular genres for MMORPGs is the Dungeons and Dragons type fantasy world. Trolls were typically non-player characters that were plentiful and annoying. In CIB, the term troll is used to denote these types of users as in the early days they were as annoying as trolls. Hence, the use of the term. What makes this inauthentic is when trolls are used to create multiple accounts and pages that they use to create these interactions. So this violates one user, one person. Therefore, there can be multiple "people" controlled by one user. It is called a farm because these trolls, equipped with the proper tools, can simulate more people in these social networks. A farm of 50 trolls can easily be working on thousands of fake accounts. All of this being controlled by a single beneficiary. This can be in the form of paid trolls or volunteer trolls.

Sock puppet is a fictitious online identity created for the purposes of deception [7]. They are generally distinguished from trolls as they are not meant to annoy or disrupt. Instead, they support activities such as creating artificial and inflated interactions, evadite platform controls and actions, and vote stacking (i.e. locally known as flying voters). Like trolls, they are also employed in farms. Although, farms with both trolls and sock puppets are generalized as troll farms.

Bots are programs that are designed to behave like real users of these social media platforms. With advances in technologies such as screen scraping and robotic process automation, programmers can create bots that behave like real people. Masking techniques can be used to ensure that platforms cannot determine that these are bots. However, unlike real people, these bots can be commanded to perform behavior that benefits the programmer such as performing interactions on the programmer's target content. Since these are programs, they can be designed to perform many more transactions than a normal human can. A possible weakness of bots and their mechanism is that they are programmed and thus their behavior can also be detected by the platforms.

Should social media platforms make it costlier or harder to create new users? These techniques exploit two characteristics of social media platforms: (1) social networks are driven by "wisdom of the crowd" behavior, and (2) social media platforms are rewarded by the market when they have more users, interactions, and content. The bigger the network the better. So social media platforms do not want to limit the number of users. Therefore, a careful balance must be struck.

EXPLOITING NETWORK EFFECTS: THE BIGGER THE BETTER

To give us a sense of the scale of the problem, Facebook took down 1.6 billion fake accounts in Q1 2022 [9]. This is a lot considering Facebook has 2.9 billion Monthly Active Users in the same quarter [10]. This is just one social media platform.

Researchers have used various mechanisms to detect these spreaders of disinformation by studying their network relationships. The use of network effects is also used to study this behavior. CIB is one of these mechanisms used by the major social media networks and independent researchers.

Independent researchers can study these campaigns using social media monitoring tools such as CrowdTangle. Information obtained from these social media monitoring systems provides information on the page or account as well as interaction activity. Social media monitoring platforms such as CrowdTangle also provide us with a historical view of these pages, accounts, interactions, and relationships [10].

This information allows us to perform algorithms to detect these cases of CIB. An example algorithm used to detect CIB would involve looking at media content posted in various social media networks, pages, and accounts to determine coordinated activity. Media content appearing across multiple pages and accounts posted within a short period of time between each other clearly exhibit a form of coordinated behavior. If these are not legitimate media sites and affiliates, then they are strong candidates for CIB. Another mechanism for detecting CIB is the use of the properties of the suspect pages and accounts instead of their relationships. This can involve looking at parameters such as creation dates. Pages and accounts created within a short period of time can display a degree of coordination that could be inauthentic. These mechanisms were applied to a study looking at social media use and CIB during the Philippine National and Local Elections of 2019 [11]. An updated study was also performed for the Philippine National and Local Elections of 2022.

COORDINATED BEHAVIOR IN PHILIPPINE ELECTIONS

The Philippines, a festive democracy of over 110 million people, has a major national election and local elections every three years. In the last electoral exercise of 2022, over 65 million voters will troop to over 37,000 voting centers nationwide to select amongst 45,000 candidates or political parties to fill over 18,000 positions (i.e. President, Vice President, Senator, Congressman, Governor, Mayor, and others). This is the most important job fair in the country and the hiring managers are the Filipino people.

This country of 110 million people is also a very well-connected populace with 82.4% of people (92 million) having social media access [12]. This shows that more

Filipinos have social media access than the number of eligible voters. Facebook, in particular, is used by 24% of Filipino adults as their primary news source. In this post pandemic survey [13], 70% of Filipino adults feel that fake news is a problem and 51% find it hard to discern fake news. That is a large number of people who are susceptible to disinformation and misinformation.

The most common mechanisms for detecting disinformation and misinformation are the use of techniques such as location-based correlation, temporal-based correlation, metadata exploration, social graphs, and content analysis. In the study of CIB in the Philippine elections, we focus on temporal-based correlation, social graphs, and content analysis. In particular, we are looking at content similarity.

In the study, two periods from February to May of the election year were studied. CrowdTangle was used to extract Facebook pages and groups for related information that could be extracted using its search Application Programming Interface (API). The key words used for both studies are "elections, halalan, ppcrv, comelec, namfrel". The studies only cover pages whose country administrators are configured to the Philippines.

In Table 6.1, we can see that there are more posts that are covered by election-related topics in the latest election. There is over a 350% increase in the number of posts, and a 134% increase in the number of accounts covered by this study.

The Fruchterman-Reingold force-directed algorithm was used to plot the relationships between these highly coordinated posts. This algorithm allows the generation of an image that allows proper spacing for the best visualization. Figure 6.1 shows an example of this image used to detect coordinated behavior in the Philippine National and Local Elections of 2022 (NLE 2022). In this image, you can see groups of highly coordinated posting of content using temporal, graph, and content analysis. Some groups are unfiltered media entities like a group of radio stations in the bottom right and the hospital group in the left side. There are also church groups clustered towards the right. There is also a government cluster with the Department of Education at the right most side. These groups are generally not political but have been sharing election-related content during this season. The content being shared can also be divided along partisan lines. The middle is interesting. You have the GO

TABLE 6.1
CrowdTangle Dataset from NLE 2019 and 2022

	2019 Study	2022 Study	Difference
Number of Posts Returned	72,320	325,490	350.07%
Number of Unique Accounts (By ID)	7,927	18,527	133.72%
Number of Posts With Shared Media	68,437	317,242	363.55%
Verified Accounts	549	1,028	87.25%
Frequently Shared URLs	4,682	18,251	289.81%
Media Elements with Coordination (Raw)	1,826	2,413	32.15%
Country of Poster	Philippines	Philippines	-

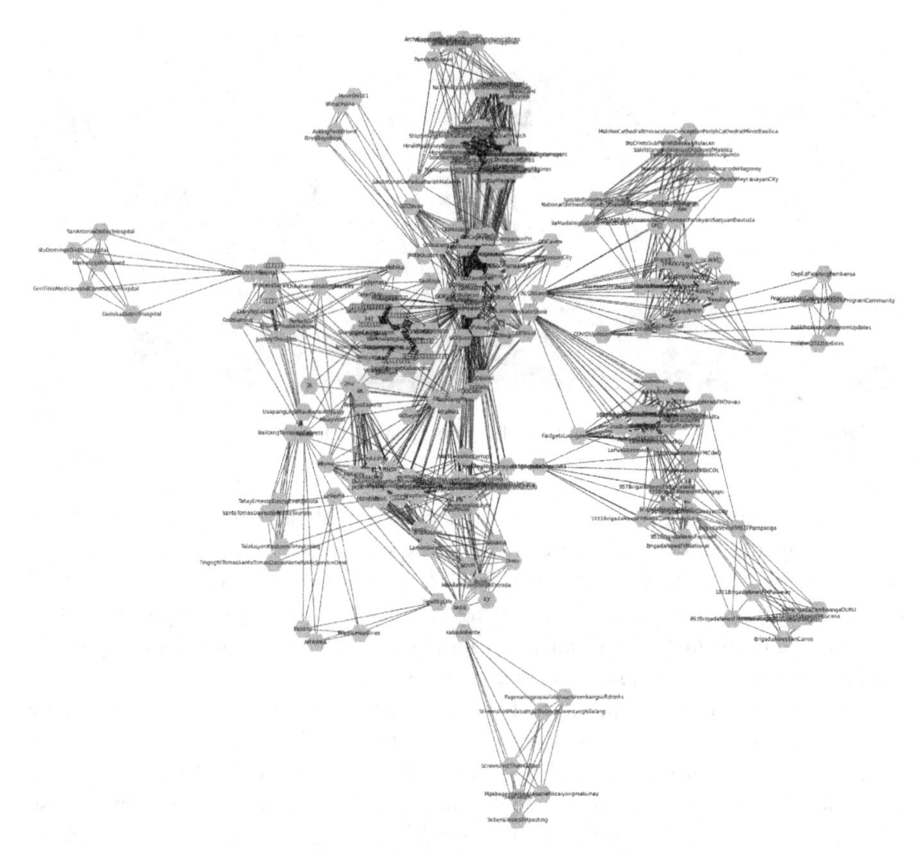

FIGURE 6.1 Coordination Graph for NLE 2022.

Philippine (advocacy group) cluster in the middle which is well connected with the issues and advocacy pages at the top and the political/candidate pages at the bottom.

In this study, coordination is defined as six or more posts sharing the same media content within one minute. This is a high degree of coordination. This number can be reduced to see more interactions and to ensure that we can observe clustered relationship network elements generated visually.

In this study, signs of coordination in social media do exist in both elections. Even political pages of competing candidates show a degree of coordination and clustering. Government and public service pages have a high degree of coordination (including other interest groups like GO Philippines). This could be a sign of political machinery or strong advocacies being used. In some cases, it is clear that some of these pages were created for specific issues or candidates. Upon looking deeper into the shared content, a number of them have already been taken down. In some cases, some of these pages have also been taken down. This could be a sign of misinformation or disinformation as platforms have already acted on them or operators have wound down their operations.

FIGURE 6.2 Sample of content with highly coordinated sharing.

Content is not simply faked. Most of them have already been taken down by the social media platforms themselves. One expected observation when looking at content is the use of honey pot material. Here some of the materials are traditional content used to bait readers into accessing the material such as sexually suggestive content, humorous material, promises of money or goods, discounts from popular e-commerce portals and websites, and many others.

A key observation is connections between opposing candidates and issues. Same operator working on both sides. This validates the view [15] that the goal of disinformation and misinformation campaigns is not always political, such as promoting a candidate. It is about dividing people along the lines of key issues. Once these divisions are created, it is possible to create highly targeted messaging to each of these groups. The users would also use content against the beliefs of the group to fan further discord. This is much more effective. This explains the high degree of coordinated action observed. See Figure 6.2.

The studies also show that social media has a prominent role here. There is an over a 350% increase in the number of social media, a 289% increase in media content with coordinated behavior, and over an 87% jump in verified accounts traffic covering the election keywords. For both study periods, it is clear that there is a good deal of coordinated behavior with a 32.15% increase in media elements that have strong coordination between the two elections. This means that social media and coordinated behavior have become more important. But, the increase in verified accounts could also show the attempt at legitimizing this coordinated behavior. This is good in a sense that auditability is possible when accounts are verified. However, it can still be used as an anchor page when using fake accounts.

FIGHTING FIRE WITH FIRE

The ability to detect CIB using social media monitoring platforms can be a great tool in fighting misinformation and disinformation campaigns allowing us to create a safer Internet. We have demonstrated its use in detecting the pages and groups showing a high degree of coordination and further looking at the content and disposition of these pages and groups to glean their authenticity. However, this is just the start.

There are limits to the above approach. First off, the focus was mainly Facebook information. This is because the Philippines is predominantly a Facebook country. Even globally, Facebook, TikTok, and Twitter get all the attention with respect to discussions on disinformation and misinformation. However, there are other platforms such as Wikipedia, YouTube, and Quora that are increasingly becoming more relevant battlegrounds in this information war. Although, TikTok seems to be giving Facebook some stiff competition. Better insight can be obtained if various social media platforms outside of just Facebook are looked across. The more data in a network the better. In addition to other social media platforms, it would also be good to have a content database, which already tags disinformation and misinformation. This allows us to use greater automation and possibly even make analysis in real time.

The second limitation is that the study was done mainly with pages and public groups only. In order to enhance the current open analysis done by the academe, it is also better to study individual accounts and not just pages and public groups. The key limitation of third-party academic studies is that we do not get individual account information due to privacy reasons. So studies are generally limited to non-privacy sensitive information, unless you were Cambridge Analytica in the early days. I am not sure the world will look too kindly on the fighting crime with crime approach. Getting individual account information allows a better view of the network. It also allows us to audit the mechanisms in place by social media platforms [16]. Are they really removing these accounts? How fast and how effective are their mechanisms? Access to this goes a long way in building trust in an ecosystem filled with distrust.

The third limitation is that studies are done using limited datasets to improve performance. Large systems studying data could be used to get deeper analysis. There are also many algorithms and methods that could be applied and tested. Many of these studies are ongoing right now.

But do not fear. There have been many developments in the ecosystem since this was brought to the forefront of public consciousness.

Constant monitoring is also necessary. A cross platform monitoring system can also help detect these types of behavior. A blend of human instinct and the best of technology. A possible way forward is the development of a scoring mechanism for pages and accounts to measure their authenticity. This inauthentic scorecard can be used to warn users of potential disinformation and misinformation. A working scorecard allows experts and people to make judgements for themselves. This is another way of using the wisdom of the crowd. Give people more information to make informed choices.

Longer term studies with historical interaction activity information can also be used to study the impact of these disinformation campaigns brought about by accounts and pages exhibiting unwarranted behavior. Studying behavior with other themes and topics are also important. This is not limited to elections after all.

However, technology is not the only weapon we can use in the fight against disinformation and misinformation. Root incentives and the ecosystem can be studied to look into other mechanisms. For example, strong identity alignment and know-your-customer (KYC) rules allow platforms to strongly validate and tie identities. This allows the platforms to create barriers to fake account creation, which is needed for troll farms to properly operate. Add cost to the step. However, care must be taken to ensure that civil liberties are not trampled on. For the platforms, they are also not incentivized to limit their numbers.

Another aspect to consider is that echo chamber creation is possible. This is because like-minded people like the same things. This is a common principle used when creating upselling and cross selling systems such as recommendation engines—Pearson's anyone? So users are grouped together with content they like. Information moving around these groups, whether true or not, will self-reinforce that particular group. Like a real echo chamber, the same sounds continue bouncing around the chamber and enhance the overall cacophony. This issue is not limited to social media but any communications media (i.e. divisiveness of US media) in general. There could be mechanisms for content curation and a degree of editorialization. However, care must be taken to ensure that it does not cross the line against free speech and creep into a form of restrictive censorship. Also, not to backtrack on the gains of an open publication environment.

WHERE DO WE GO FROM HERE?

Role of platforms. Platforms include service providers, network providers, and social media platforms. As they are in the path of the transaction, platforms have the most access to the information and data. Additionally, being in the path of the transaction gives these platforms a large amount of power to enforce some of the changes that are needed. This means platforms should continually be vigilant about forms of manipulation in the ecosystem and how people attempt to exploit these platforms. Greater transparency is also needed with respect to actions taken. There is a balance here as well. There is a concern that too much transparency may allow exploiters to know more and study the platforms. At a minimum with better transparency, users and other ecosystem stakeholders are able to manage their expectations and act accordingly.

Role of Artificial Intelligence (AI) in the future. As people get used to this level of automation and delegate more decisions to these services, this can possibly make users more susceptible to manipulation. I believe that every new generation will be willing to seed more decision making to platforms and algorithms. This is particularly true as every generation gets used to this high level of automation and curation driven by AI. Again, the call would be for a better understanding of how these algorithms drive outcomes. More dialogue on this realm of algorithmic transparency is needed.

Role of independent research. Do we just trust the platforms and make it their job? As independent researchers, we are in a position to take an "outside in" look at the situation. It also allows research into cross platform effects and how this type of manipulation makes its way to other systems. Evasion techniques can also be studied when they cross platforms. The broader academic community can also do much to strengthen studies into these forms of manipulation and be part of possible unbiased solutions.

Role of the rest of us. We have been granted a powerful communications tool that spans the globe. Everybody is a publisher. But, not all publishers play by the rules. As users, we should be aware that manipulation is possible, and we should behave accordingly. The information might be curated with the "wisdom of the crowd" but we should still exercise proper discernment when using this information.

THE PROBLEM OF FAKE NEWS IS REAL

Mainstream media entities and journalists do not like the term fake news. For after all, news must not be fake. Right? The problem is much larger and complex that can fit in a single chapter. However, we covered some key aspects of the disinformation and misinformation problem. We looked at its context and a means to fight it using CIB detection. These are just some of the many mechanisms that can be used in the battle against disinformation and misinformation. Detecting disinformation and misinformation using CIB is just one of them. But, if algorithms can be used to exploit these platforms and ecosystems then algorithms can also be used to keep these same platforms and ecosystems honest. The danger is clearly that well organized and funded groups can exploit the system to propagate their narratives. Many of which are destructive to democracy.

It is true that platform providers in this ecosystem have an outsized role in maintaining the Internet as a free and open space. Many of these social network systems are designed assuming "one person, one voice". Many users are conditioned to expect these systems to operate with "one person, one voice". However, given clear evidence of manipulation using disinformation and misinformation, platforms and users alike must be careful when looking at information from the Internet. We must be wary. Even ordinary users must be educated to be careful with such information from the Internet. There is a need for more research in this space, more awareness building, and more cooperation. It takes the cooperation of all participants in this "Internet Commons" to ensure that it remains a free, open, and healthy space for generations to come.

REFERENCES

[1] Giglietto, F., Righetti, N., Rossi, L. and Marino, G. (2020). "It takes a village to manipulate the media: Coordinated link sharing behavior during 2018 and 2019 Italian elections". *Information, Communication and Society*, 1–25.

[2] *Global Digital Overview*. https://datareportal.com/global-digital-overview. Last accessed: June 15, 2022.

[3] Lewis, L. (2021). "What happens in an Internet minute". www.allaccess.com/merge/archive/32972/infographic-what-happens-in-an-internet-minute. Last accessed: June 19, 2022.

[4] Page, L., Brin, S., Motwani, R. and Winograd, T. (1999). *The PageRank Citation Ranking: Bringing Order to the Web*. Stanford InfoLab.

[5] Ledford, J.L. (2015). *Search Engine Optimization Bible* (Vol. 584). John Wiley & Sons.

[6] Weedon, J., Nuland, W. and Stamos, A. (2017). *Information Operations and Facebook*. https://about.fb.com/news/2017/09/information-operations-update/. Last accessed: June 16, 2022.

[7] Kats, D. (2020). "Identifying sockpuppet accounts on social media platforms". www.nortonlifelock.com/blogs/norton-labs/identifying-sockpuppet-accounts-social-media. Last accessed: June 19, 2022.
[8] Facebook Community Standards Enforcement Report Q1 2022 https://transparency.fb.com/data/community-standards-enforcement/. Last accessed: June 19, 2022.
[9] Dixon, S. (2022). "Facebook: number of monthly active users worldwide 2008–2022". www.statista.com/statistics/264810/number-of-monthly-active-facebook-users-worldwide/. Last accessed: June 19, 2022.
[10] CrowdTangle Team (2020). "CrowdTangle". *Facebook*, Menlo Park, California, United States. https://apps.crowdtangle.com/admuscienceengfacebook/lists/pages. Last accessed: June 15, 2022.
[11] Yu, W. (2021). "A framework for studying coordinated behaviour applied to the 2019 Philippine Midterm Elections". Proceedings of the 6th International Congress on Information and Communication Technology (ICICT 2021). Virtual Conference, February 2021.
[12] Philippine Digital Overview. https://datareportal.com/reports/digital-2022-philippines. Last accessed: June 15, 2022.
[13] Philippine Daily Inquirer (2020). "SWS: 45% of Filipino adults use internet; 1 in 4 read news through Facebook". https://newsinfo.inquirer.net/1332871/sws-45-of-filipino-adults-use-internet-1-in-4-read-news-through-facebook. Last accessed: June 15, 2022.
[14] Philippine Daily Inquirer (2022). "70% of Pinoys say fake news a serious problem–SWS". https://newsinfo.inquirer.net/1560828/sws-70-of-pinoys-say-fake-news-a-serious-problem. Last accessed: June 15, 2022.
[15] CITS. "The danger of fake news in inflaming or suppressing social conflict". www.cits.ucsb.edu/fake-news/danger-social Last accessed on: June 19, 2022.
[16] Gleicher, N. (2020). "Removing coordinated inauthentic behavior". https://about.fb.com/news/2020/10/removing-coordinated-inauthentic-behavior-september-report/

7 Refining the Sweeney Approach on Data Privacy

Rodney H. Cooper
University of New Brunswick
Fredericton, NB, Canada

Wayne Patterson
Patterson and Associates
Washington, DC, USA

INTRODUCTION

It is increasingly clear that two phenomena have grown extensively over the past two decades: first, the exponential growth of cyberattacks in virtually every computing environment; and second, public awareness (whether accurate or not) of one's vulnerability to attacks that may be directly aimed at the individual, or more generally to an organization that maintains widespread data on the entire population.

The pioneering research by Sweeney demonstrated the vulnerability of most residents and computer users in the United States to the easily available demographic data necessary to identify sensitive information about any individual: "It was found that 87% (216 million of 248 million) of the population of the United States had reported characteristics that likely made them unique based only on a 5-digit ZIP, gender, date of birth." [1]

However conclusive Sweeney's research was concerning the citizens and residents of the United States, her research only provided a template for developing similar estimates regarding other countries throughout the world. It is our purpose to extend

DOI: 10.1201/9781003415060-9

the previous research to develop similar estimates regarding residents' vulnerability to data attacks using similar demographic data.

The estimates that can be developed are dependent on assumptions as to the distribution of the population in a given country, as well as an assumption of the number and the distribution of the postal code districts. Previous studies in this area of research have established estimates of population distribution in comparison to postal code districts.

A variation on this estimate can be done by redefining the problem as one concerning the "assignment of pigeons to pigeonholes."

"PIGEONHOLES"

Our approach will be to conduct assessments of the data for the number of persons that can fit into each of the potential categories, or "pigeonholes", in a frequently-used term in computer science. Sweeney's earlier study is to say that of all the pigeonholes, approximately 87% have no more than one datum (that is, no more than one person) assigned to that pigeonhole.

The number of pigeonholes in Sweeney's study for the United States is calculated by the product of the potential number of persons identified by birth date including year, gender, and 5-digit ZIP code. The USA 5-digit ZIP code may be described as "NNNNN", where each "N" may be represented by a single decimal digit, i.e. {0, 1, 2, …, 9}. In the case of Canada, which will be the focus of this paper, the syntax for Canada Postal Code construction is "ANA NAN", where again, the "N" refers to a decimal digit, and "A" refers to a capital letter in the set, i.e. {A, B, C, …, Z}. The cardinality of this set is 26.

The contribution to the "pigeonhole" number related to gender is 2, say $p_g = 2$. For birth date, we approximate the number of values using 365 for days of the year (a slight simplification ignoring leap years), multiplied by the number of years, estimated by the country's life expectancy in years [3]. Call this p_b. The final relevant factor in estimating the number of pigeonholes is the number of potential postal codes, p_p. Then the total number of pigeonholes is:

$$\#\text{pigeonholes} = p_g \times p_b \times p_p = 2 \times (365 \times \text{life expectancy}) \times p_p$$

One remaining problem is the calculation of the number of postal codes, p_p. It is an easy calculation to find the maximal value for p_p that we will call p_{pmax}. For example, for the 5-digit US ZIP code system, that maximal value is $p_{pmax} = 10^5 = 100{,}000$. At the time of Sweeney's research, the number of ZIP codes actually used was $p_p = 29{,}343$ ([1], page 15), or 29.3% of the total number of ZIP code values. At present, the number of ZIP codes in use is 40,933. In theory, the comparable number for Canada would be $10^3 \times 26^3 = 17{,}576{,}000$.

Given available data for all world countries, the value p_p is often not made public. The previous work in [2] established Table 7.1.

As indicated above, the previous research by Sweeney showed that approximately 87% of US persons could be uniquely identified by the criteria described above. The above table shows a consistency in these results in that the value of

TABLE 7.1
Average Number of Residents per Sweeney Criterion (Pigeonholes) for a Sample of Countries (Data as of 2019)

Country	Gender x Days/yr	Life Expectancy	Actual Postal Codes	Pigeonholes	Population	Average No. of Persons/ Pigeonhole (APP)
Great Britain	730	81.2	1,700,000	1.008E+11	6.62E+07	0.0007
Canada	730	82.2	834,000	5.005E+10	3.66E+07	0.0007
South Korea	730	82.3	63,000	3.785E+09	5.10E+07	0.0135
Spain	730	82.8	56,542	3.418E+09	4.64E+07	0.0136
Mexico	730	76.7	100,000	5.599E+09	1.29E+08	0.0231
China	730	76.1	860,000	4.778E+10	1.41E+09	0.0295
Poland	730	77.5	21,965	1.243E+09	3.82E+07	0.0307
Japan	730	83.7	64,586	3.946E+09	1.27E+08	0.0323
Israel	730	82.5	3,191	1.922E+08	8.32E+06	0.0433
France	730	82.4	20,413	1.228E+09	6.50E+07	0.0529
Netherlands	730	81.9	5,314	3.177E+08	1.70E+07	0.0536
Russia	730	70.5	43,538	2.241E+09	1.44E+08	0.0643
United States	730	79.3	40,933	2.370E+09	3.24E+08	0.1369

average persons/pigeonhole (APP) for the United States is approximately 13.7%; in other words, since APP = 13.7%, the percentage of "pigeonholes" with only one entry is 1–APP ≅ 87%. This coincides with the Sweeney research that the percentage of Americans who can be identified uniquely by the Sweeney method, or by pigeonholes, is about 87%.

For the purposes of this study, we will consider an analysis of one particular secondary level of government, namely in reference to the national data for Canada, to analyze the second level of government in Canada, which consists of 10 provinces and three territories. The following table will demonstrate the comparable results to Table 7.1 for the selection of countries to an analysis of the Canadian data, not only for the nation as a whole, but for the provinces and territories considered separately. This data is described in Table 7.2.

To describe Canadian postal codes, we will first note that the overall syntax with such code is of the form "ANA NAN", where, as before, "A" refers to an alphabetic character in {A … Z}, and "N" refers to a decimal digit {0 … 9}. We will also describe Canadian postal codes as having a prefix "ANA" and a suffix "NAN."

We consult [4], [5] to determine the number of prefixes in these codes by Canadian province and territory and for Canada as a whole, as well as life expectancy. We estimate the number of suffixes by computing all possible values for the suffix form "NAN", in other words 10 × 26 × 10 = 2,600, which we call "suffixes."

The number of pigeonholes from the above is calculated by:

$$\text{Pigeonholes} = \text{GD} \times \text{Life expectancy} \times \text{Postal Codes Prefixes} \times \text{Suffixes.}$$

TABLE 7.2
Average Number of Residents per Sweeney Criterion (Pigeonholes) for a Canadian Provinces and Territories, Sorted by the Value APP (Data as of 2022)

Province/Territory	GxD	Life Expectancy	Actual Postal Codes	Pigeonholes	Population	Average Persons/ Pigeonhole (APP)
North West Territories	730	77.4	1	56,502	45,607	0.80717
Ontario	730	82.4	471	28,331,592	15,007,816	0.52972
Alberta	730	81.6	156	9,292,608	4,500,817	0.48434
British Columbia	730	82.4	193	11,609,336	5,286,528	0.45537
Prince Edward Island	730	81.6	7	416,976	167,680	0.40213
Saskatchewan	730	80.3	53	3,106,807	1,186,308	0.38184
Manitoba	730	80.1	70	4,093,110	1,393,179	0.34037
Quebec	730	82.9	433	26,203,861	8,653,184	0.33023
Newfoundland and Labrador	730	80	35	2,044,000	522,875	0.25581
Yukon	730	79	3	173,010	43,249	0.24998
Nova Scotia	730	80.4	77	4,519,284	1,007,049	0.22283
Nunavut	730	71.1	5	259,515	.840,103	0.15453
New Brunswick	730	80.7	111	6,539,121	800,243	0.12238
Canada	730	81.8	1615	96,438,110	37,854,395	0.39252

For example, for Alberta, this value is:

$$730 \times 81.6 \times 156 \times 2600 = 9,292,608.$$

We order the complete table for the provinces and territories in decreasing value of the value of APP, as shown in Table 7.2.

Examining this table allows us to conclude that it is harder to identify an individual value in only the jurisdictions Northwest Territories, Ontario, Alberta, British Columbia, and Prince Edward Island. For Canada as a whole, approximately 60.7% (= 1–39.3%) of individuals may not be individually determined.

In the particular case of Canadian postal data, it is difficult to draw comparable conclusions as can be determined in many other countries. This is because there is such a large number of available postal codes, so in the vast majority of cases, this "pigeonhole" criterion will rarely allow for more than one entry per pigeonhole; in other words close to 100% of Canadians can be identified by the Sweeney criteria. This is indicated in Table 7.1, with only Great Britain and Canada among the countries analyzed demonstrating this phenomenon. However, this analysis allows for a qualitative differentiation of the ease of detecting persons by the stated criteria in the various Canadian jurisdictions.

The same analysis applied to other countries considered would result in more quantitative results for the secondary level of jurisdiction.

REFERENCES

[1] Sweeney, L., *Simple Demographics Often Identify People Uniquely.* Carnegie Mellon University, Data Privacy Working Paper 3. Pittsburgh (2000).

[2] Patterson, W. and Winston-Proctor, C., *An International Extension of Sweeney's Data Privacy Research.* Advances in Human Factors in Cybersecurity, Proceedings of the AHFE 2019 International Conference on Human Factors in Cybersecurity, July 24–28, 2019, Washington D.C., USA, pp. 28–37.

[3] Wikipedia. *List of countries by life expectancy* https://en.m.wikipedia.org/wiki/List_of_countries_by_life_expectancy

[4] World Postal Codes. https://worldpostalcode.com/canada/

[5] Wikipedia. *List of Canadian provinces and territories by life expectancy* https://en.m.wikipedia.org/wiki/List_of_Canadian_provinces_and_territories_by_life_expectancy

Section III

Cybersecurity Concerns in the Home and Work Environment

8 Cybersecurity Hygiene
Blending Home and Work Computing

Thomas W. Morris and Jeremiah D. Still
Department of Psychology, Old Dominion University
Norfolk, VA, USA

CYBERSECURITY HYGIENE: BLENDING HOME AND WORK COMPUTING

Overview

Remote work demand is increasing, and how it is performed is changing. The hybrid home-work computing environment is becoming a soft target for potential cyberattacks. This chapter examines cybersecurity hygiene within the rapidly evolving hybrid home-work computing environment. The concepts of cyber hygiene and the hybrid home-work computing environment are explored and abstractly defined. A reflection on cybersecurity and physical security reveals unique security knowledge expected to be elicited from home computer users in the cyber environment. We explain how the learning of cybersecurity behaviors is different from physical security behaviors due to the lack of cultural transfer between generations. Key differences between the home and work computing domains are described. Cybersecurity experts have touted many possible solutions to the threats created by hybrid home-work computing. However, even with these solutions, it is impossible to design out the human element. Users need to be a critical part of the cybersecurity informational loop. This chapter highlights the needed behavioral interventions to ensure users are part of this process. And, discusses how training, nudges, warnings, and better visualizations are needed to meaningfully mitigate cyber risk. An initial list

DOI: 10.1201/9781003415060-11

of unique hygiene recommendations for users in the hybrid home-work computing environment is provided. Finally, we discuss the need for future research in the hybrid home-work computing domain.

Recent History

According to the Owl Labs 2021 State of Remote Work survey, it is estimated that globally 16% of companies employ a fully remote workforce. They also found that 62% of workers aged 22 to 65 claim to work remotely occasionally. As the cyber landscape evolves, home and work computing environments are becoming more intertwined. Unfortunately, home computing cybersecurity behaviors have been studied less than organizational cybersecurity behaviors. This is due to the organizational infrastructure of information technology (IT) professionals, best practices, and reporting standards (Howe et al., 2012). As home and work computing become more entwined, the gap in our understanding of home computing (HC) cybersecurity behaviors will become more costly. The recent abrupt change from in-office to remote work due to the COVID-19 pandemic has increased our need to better understand HC users' cybersecurity behaviors. This perfect storm of rapid adaptation of telework technologies and the lack of research regarding how to best support HC users in the telework environment will undoubtedly result in new opportunities for bad actors.

Poor cybersecurity behaviors are costing companies millions of dollars. For instance, in a 2021 ransomware attack on a US oil pipeline system, hackers demanded $4.4 million to unencrypt the files needed to continue the pipeline operations. This attack on critical infrastructure caused the pipeline to halt operations. To resolve this crisis, the pipeline company paid the attackers' ransom. But even after paying the ransom, it took a long time for the files to decrypt, which led to a significant gas shortage on the east coast of the United States. It was later discovered that the attackers gained access to the system through a Virtual Private Network (VPN) because a single remote employee had a weak password. This attack shows the importance of basic cyber hygiene for teleworkers.

CYBERSECURITY HYGIENE

"Cyber" hygiene as a concept feels familiar but has not been fully defined in the cybersecurity literature. Much research is underway regarding the operationalization of cyber hygiene as a concept. For instance, one definition describes cyber hygiene as "the cyber security practices that online consumers should engage in to protect the safety and integrity of their personal information on their Internet-enabled devices from being compromised in a cyber-attack" (Vishwanath et al., 2020, p. 2). Another perspective defines cyber hygiene as personal cybersecurity (Clemente, 2021). Interestingly, there has been discussion about changing the name from cyber hygiene to OPSEC (operational security) to align with other fields of study (Vishwanath, 2021).

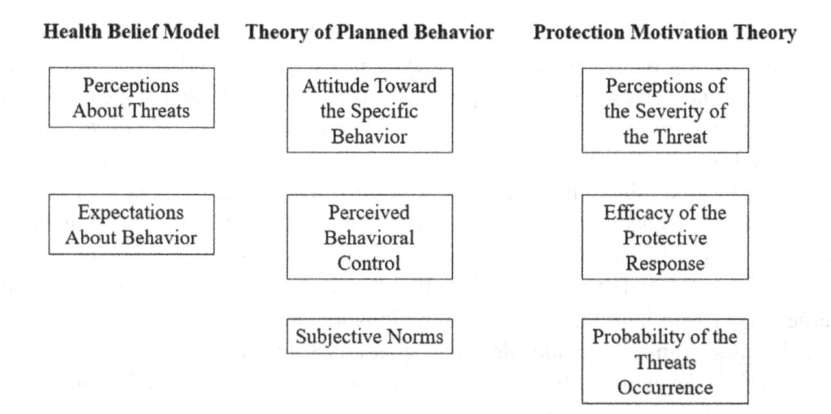

FIGURE 8.1 Main Factors Considered as the Determinants of Preventative Health Behaviors.

***Note.* Source: Dodel & Mesch (2017) and Howe et al. (2012).**

Historically, the concept of cyber hygiene stems from social cognitive theories attempting to explain risky behavior in preventative health. For example, Detweiler et al. (1999) found that when messaging highlights the potential gains of using sunscreen then people are significantly more likely to obtain and use sunscreen when compared to messages that highlight the protentional losses from not using sunscreen. This study aimed to promote the health prevention behavior of using sunscreen but the results have been successfully applied to many different domains, including messaging in cybersecurity.

Traditionally health-focused researchers have been able to successfully translate health-belief models, theories of planned behavior, and protection motivation theories into the cybersecurity domain (Dodel & Mesch, 2017). Figure 8.1 provides an overview of the main factors considered as the determents of preventative health behaviors.

In each of these models, sociodemographic variables and environmental cues are considered. The *health-belief model* attempts to identify the relationships between specific variables and the likelihood of taking preventative health action (American Psychological Association, n.d.). The *theory of planned behavior* posits that the intention to perform a behavior is the main indicator of a specific behavior (Howe et al., 2012). *Protection motivation theory* describes how individuals are motivated to react in self-protective ways towards a perceived threat (van Bavel et al., 2019). While some researchers argue that the concept of cyber hygiene should be moved from the health literature, a significant amount of past research confirms that these preventative health theories apply to the cybersecurity domain and can provide some explanation for cybersecurity hygiene behaviors (Li et al., 2019; Dodel & Mesch, 2017; Howe et al., 2012). Clearly, the health prevention literature offers valuable insight into how to predict and promote cyber hygiene. Therefore, even if the name evolves from the health literature, we must not throw the baby out with the bathwater.

Now let us turn our attention to the concept of cyberspace itself. *Cyber* is defined as communication over any IT system that stores, retrieves, and sends information (c.f., Lexico, n.d.). However, many people would not believe that to be true to the modern conceptualizations of the Internet. As a caveat, many people do not believe that this definition of cyber is a good representation. For example, when someone is using a landline telephone, they send and retrieve information, but do not claim to be using cyberspace. This distinction between whether information technologies belong to cyberspace can become controversial, especially when introducing services like Voice over IP (VoIP). Historically called IP telephony, VoIP services require the Internet to perform landline telephone functions making this device a part of cyberspace. We first want to consider defining what constitutes cyber and what does not. Understanding the evolving boundaries of cyberspace is necessary for forming a good operational definition of cyber hygiene.

HYBRID HOME-WORK COMPUTING

Defining cyberspace boundaries presents unique challenges when considering hybrid home-work computing environments. We define the *hybrid home-work computing environment* as the cyberspace emergent from a combination of both a user's HC environment and work computing environment. The emergent cyberspace brings some of the characteristics and properties of both computing environments. For example, a hybrid home-work computing environment incorporates both home and work resources (e.g., networks and services). Figure 8.2 provides a visual conceptualization of this integration between personal and organizational computing.

These hybrid home-work environments are rapidly expanding in complexity. With this expansion comes many new questions that need to be answered. Can you separate an individual's private HC environment from their professional computing environment? Should it be temporally divided between being on or off the clock? Alternatively, resource consumption-based, whenever an employee uses a company device or service. It is simple to argue that a user is in the work environment anytime they are connected to company resources and the HC environment at all other times. However, networks do not stop capturing information based on these factors. An HC user who torrents a video game, for example, may do this when they are not actively connected to their work network, but the malware that is introduced to the HC network will potentially be spread to their work network the next time this user connects to their work resources, such as VPN services. In both the physical and cyber sense, the hybrid home-work computing environment does not have the inherent separation between space as the traditional work computing environment. As a result of this lack of separation, further precautious and mitigation techniques will need to be implemented to ensure security and privacy. Learning how to secure this new hybrid home-work environment will be difficult without additional training and interface transparency that supports greater situational awareness.

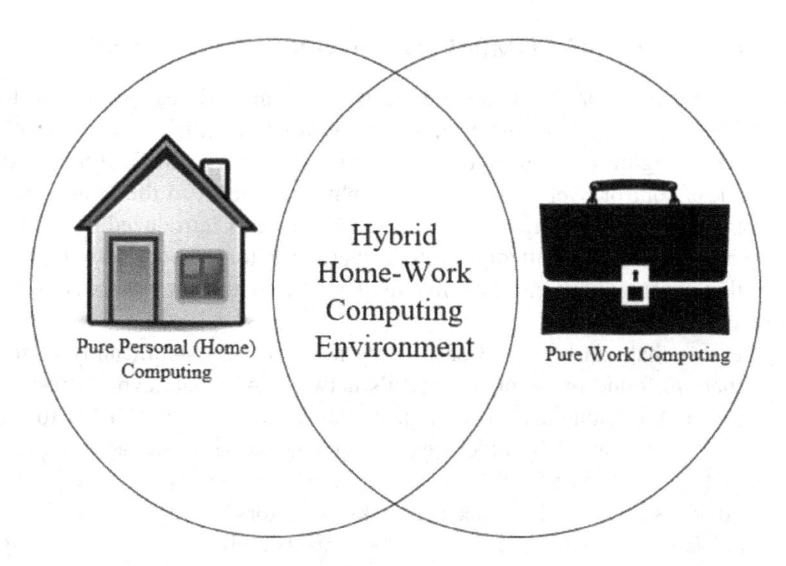

FIGURE 8.2 Conceptualization of the Hybrid Home-Work Computing Environment.

***Note.* The hybrid home-work computing environment integrates characteristics from both home and organizational computing environments.**

HOW IS CYBERSECURITY ANALOGOUS TO PHYSICAL SECURITY?

We can expect users to learn cyber hygiene behaviors like other previously taught security behaviors. As children, people are often introduced to the general concept of security first through their parents or caregivers. Parents teach their children ways to safeguard themselves from intruders (i.e., lock the doors). Unfortunately, this kind of innate training rarely happens in the cyber world as our landscape is rapidly expanding. Currently, parents lack good cyber hygiene themselves. Unfortunately, they are implicitly passing on poor cybersecurity behaviors to their children. Therefore, we need to support users with better interface transparency as they lack the protective capabilities that are almost second nature in physical safety.

People often do not perceive security in cyberspace as they do in the physical space, which leads to risky behaviors that would seem out of place in a physical environment. For instance, most people intuitively know not to leave valuable personal information in a shared public space. It is abnormal for people to leave their birth certificate or social security card in an easily accessed location. This same intuitive protective behavior does not appear to be apparent in cyberspace. People readily save sensitive information on unsecured devices that are easily accessible. Regardless of users' understanding of cyberspace security, they are now asked to maintain both their home and work networks and associated services.

HOME COMPUTING COMPARED TO WORK COMPUTING

The home network provides unique security vulnerabilities compared to an organization's network. For instance, organizations have little to no control over the network configuration or devices connected to personal networks. Organizations also have less control over those who have physical access to the home network. Some companies are working to address potential threats introduced by the hybrid home-work network. They attempt to intensively monitor network activity, but it is unlikely that organizations can establish the level of control they possess over their networks.

The devices that can be found on a home network vary significantly compared to those that are found on an organization's network. Adequate cyberinfrastructure maintenance is commonplace on organizational networks, which helps to ensure that vulnerabilities caused by older devices and out-of-date software are properly managed. Moreover, Internet of Things devices are common on the home network, and these devices present unique security risks as vectors for attack by threat actors. *Internet of Things* (IoT) devices are everyday objects embedded with technology to allow smart features. Because these are commercial products, the device's usability takes precedence over security requirements (Abomhara & Køien, 2015). According to Westscott (2019), the average home computer user has 11 IoT devices connected to their home network. The number of IoT devices connected to home networks is only growing, making home networks more challenging to protect.

There are also less formalized cybersecurity incident reporting practices in the HC environment compared to the work computing environment. When a user encounters a potentially malicious website or message in the work environment, it is commonplace to report this to its IT department. The same cannot be said in the HC environment. Often, people have nowhere to report malicious cyber encounters at home. Further, no warning messages about ongoing attacks are being sent out. This leaves the home network vulnerable to attack and contributes to the lack of situational awareness.

The *CIA Triad* security model highlights differences in work and HC goals. The model characterizes the role of cybersecurity as protecting the confidentiality, integrity, and availability of data. It is impossible to maximize all three of these components (Clemente, 2021). Often HC users are willing to trade privacy (i.e., confidentiality) for the availability of information, but companies are willing to pay to maintain the privacy of their information. Due to the conflict between the cybersecurity goals of the HC user and an organization, the hybrid home-work network suffers. This highlights the need to go beyond technological solutions. Behavioral interventions will need to be employed to improve the overall cyber hygiene of end-users and strengthen this weak point in the hybrid home-work computing system.

The *home computer user* is a novice who has no formal training in using computing yet employs technology to accomplish everyday tasks (Howe et al., 2012). HC users vary significantly in demographics, activities, and motivations, which makes the general cybersecurity habits of these users particularly difficult to predict. Past discussions on HC users are generally limited to users in the strictly home environment; therefore, only gleam part of the security landscape. For example, an employee may receive cybersecurity training before being provided remote access to an organization's network, but others in the user's household do not receive training. This could result in accidental or unauthorized access to organizational resources. This highlights the need for overall improvement of the HC user's cybersecurity hygiene and the need to introduce design solutions that go beyond training.

SOLUTIONS TO PROBLEMS IN THE HYBRID HOME-WORK COMPUTING ENVIRONMENT

We start by conceptualizing some of the initial solutions. Organizations might attempt to resolve many emerging concerns by becoming entirely dependent on cloud computing services. Nevertheless, users will still face various threats, including cryptocurrency mining abuse, phishing campaigns, and ransomware (Google Cybersecurity Action Team, 2021). Due to the relative novelty of cloud computing, the implementation of services themselves presents a unique security risk since they lack basic controls to prevent end-users from engaging in potentially malicious activities (Google Cybersecurity Action Team, 2021).

Another solution being considered is to set up a separate network in people's homes to use for only work. While this measure is impractical for most companies due to the cost, even this drastic solution does address the human element. Users will be expected to switch back and forth between their home and work networks when accomplishing daily tasks. Network engineers have worked diligently to create a streamlined, seamless exchange of information. A loss in network identity salience can produce unintended outcomes like HC users accidentally employing computing resources to accomplish personal tasks. We may need to design for network salience to ensure the necessary transparency to discriminate between work and personal resources. These updated design elements could include visualizations to remind the user of which network is currently connected. Another potential issue is that there are many different users within the home environment. This could include guests, roommates, family members, and any other person that might enter the home. Unlike in an organization, there exists little monitoring of the physical environment where the network devices are stored making homes soft targets. Regardless of the solution, behavioral interventions must also be in play to effectively mitigate cyber threats. A popular way to encourage decision making is to employ a nudge at critical decision points during system interactions.

ROLE OF BEHAVIORAL INTERVENTIONS TO IMPROVE CYBER HYGIENE

Nudges can predictably influence HC users' decision-making process (Caraban et al., 2019). From an information processing perspective, these nudges can be reflective to the user (e.g., notifications) or automatic (e.g., default settings). And, can vary in the level of transparency provided to the user (Caraban et al., 2019). Past research has shown that notification nudges designed using protection motivation theory improve cyber hygiene more than notifications that rely solely on fear appeals (van Bavel et al., 2019). It was shown that sociodemographic factors and risk attitudes influenced cyber hygiene behaviors. These findings support the application of the protection motivation theory. It appears that well-designed nudges can effectively improve cyber hygiene.

While nudge interventions help users make better cyber hygiene choices, a *warning* alerts users when a potentially dangerous action has occurred or when danger is imminent. Warnings help guide users away from potentially dangerous behaviors. Critically, they provide users with an informational feedback loop, allowing HC users to associate their poor cyber hygiene practices with consequences. Unfortunately, users have difficulty understanding warnings in cyberspace (Xiong et al., 2018). HC users lack comprehension of warnings even when purposely designed to best communicate risk (Yang et al., 2017). Warnings play an essential role as a last-ditch effort to prevent users from making poor cyber hygiene decisions. Future research needs to develop the algorithms that generate the warning and the design research to explore user compliance.

Training has been shown to reduce the number of cybersecurity incidents within organizations (Kweon et al., 2019). Lessons from the aviation training literature can help us appreciate the need for cybersecurity training. Two Boeing 737 Max airplane crashes in 2019 resulted from a lack of training (Campbell, 2019). The pilots were new to the automation feature meant to help keep the nose of the plane level. The training on this new feature was considered too costly. And, it would ground pilots from flying the new plane until it was completed. In other words, training employees on critical safety would negatively impact the airline's limited workforce too much. As a result, the Boeing 737 Max pilots were untrained on how to override the new automated system. This desire to avoid training employees on safety and the overreliance on automation eventually led to the two deadly crashes killing 346 people. These tragedies exemplify the struggle of re-training users every time there is a system change. Cybersecurity training can be costly for organizations, but it can be even more costly for organizations to not properly train their employees as cybersecurity automation is being brought to critical infrastructures, such as fuel pipelines and hospital systems.

It is unlikely that future cybersecurity problems will be automated away. Instead, HC users will be asked to partner with automated systems. Users must understand the strengths and limitations of cybersecurity automation. If not, overreliance and misunderstanding will produce vulnerabilities that might produce high costs. Partnering with automated cybersecurity systems is currently complex and needs further research and development. However, we at least need HC users to make good cyber hygiene decisions. Researchers have shown that current cybersecurity training and awareness campaigns are ineffective at improving cyber hygiene (Cain et al., 2018). Often cybersecurity training attempts to address corporate legal accountability instead of effective instructional design. Clearly, the current instruction needs to be re-evaluated to determine the factors that best improve cyber hygiene learning. For example, it is not enough to just scare people through fear appeals into making good cybersecurity decisions (Dupuis et al., 2021). Instead, users must be informed of the nature of cyber threats for cyber hygiene behaviors to show improvement. A well-designed cyber hygiene training will arm HC users with the knowledge to understand threats and better support users when faced with important decisions.

No single behavioral intervention is shown to help users make better cyber hygiene decisions. Developers need to employ a combination of nudges, warnings, and training materials. Further, a serious attempt needs to be made to improve the visualization and usability of security-related systems. Only a holistic approach will help users and harden our critical systems. For instance, it was shown that combining a warning with training helped participants better long-term cybersecurity decisions (Xiong et al., 2018). Calls for human-centered design work in cybersecurity have been made (Still, 2016). The cybersecurity interfaces of the future need to be more intuitive and transparent to assist users in making timely and appropriate hygienic cyber decisions. It is well known that interactions can vary along a continuum of intuitiveness (Still et al., 2014). Due to this, designers need to prioritize ease of use by attempting to maximize the intuitiveness of cybersecurity systems. In tandem with other behavioral interventions, the development of intuitive interactions in the hybrid home-work computing environment will help to better support users.

CYBER HYGIENE TIPS FOR THE HYBRID HOME-WORK COMPUTING ENVIRONMENT

Clemente (2021) offers many best practices to help users maintain good cyber hygiene in the HC environment. We offer five formal recommendations for users to maintain good cyber hygiene in the hybrid home-work network.

1. *Separate work and personal devices.* It is crucial that computing be separated between work and personal use due to the different computing priorities in each domain. Work computing devices should solely be used for work.
2. *Secure access to devices.* Devices used for work in the home should be secured with a strong password and placed in a locked room when not in use.
3. *Encrypt devices.* All connected devices must be protected with robust and up-to-date encryption. This will help to prevent the chance of unauthorized access and exposure to private organizational information.
4. *Ensure that protected data is stored correctly.* Be aware of data storage best practices for your profession. Know where to store business files to help prevent leaks from work systems into the home system.
5. *Regularly partake in cybersecurity awareness training.* Work to maintain situational awareness of cybersecurity threats and learn about new preventative measures. Aim to create a culture of teamwork and continuous learning around these topics.

This list of recommendations is far from comprehensive but serves as a starting point for organizations to consider as teleworking demand increases. Further recommendations will need to be made that are specific to the unique needs of the hybrid home-work computing environment.

FUTURE RESEARCH

We have a dire need to develop a more robust definition of cyber hygiene, or at the very least decide on a multidisciplinary term that can be used for translational research between disciplines. We need a richer representation of the typical HC users' mental models. What do novice users know, and how can training, nudges, warning, and visualization support cyber hygiene maturity? More research is needed to determine the unique technical and behavioral threats presented in the rapidly changing hybrid home-work computing environment. Hopefully, future research will provide well-defined best practices or specialized behavioral interventions. There have already been calls to further develop best practices for IoT devices and establish human factors programs to help users mitigate cybersecurity risks (Momenzadeh et al., 2020; Nobels, 2019). We hope this chapter draws much-needed attention to the hybrid home-work computing environment and further adds to the ongoing list of research that must be addressed.

CONCLUSIONS

Cybersecurity hygiene is not well defined. There is still no universally accepted operational definition of this concept, which results in varying ways of measuring cyber hygiene. As the home and work computing environments merge to create a growing hybrid home-work computing landscape, the need to better operationalize (and hopefully one day be able to predict) the cyber hygiene of users becomes increasingly apparent. The concept of telework itself is not new, but the landscape around this practice continues to evolve rapidly. It is also clear that behavioral interventions and intuitive design will play an important role in supporting users. These interventions will need to be designed to increase the intention, motivation, and knowledge of good cyber hygiene practices in home-work computer users. Future work should aim to operationalize cyber hygiene as a concept and develop best practices for hybrid home-work computing.

GLOSSARY

CIA Triad – cybersecurity model that characterizes it as protecting the confidentiality, integrity, and availability of data.

Cyber Hygiene – the cyber security practices that online consumers should engage in to protect the safety and integrity of their personal information on their Internet-enabled devices from being compromised in a cyberattack.

Cyber – relating to or characteristic of the culture of computers, information technology, and virtual reality.
Health-Belief Model – attempts to identify the relationships between specific variables and the likelihood of taking preventative health action.
Home Computer User – characterized as a novice that has no formal training in the use of a computer but uses computing devices in the home environment to accomplish tasks to support their lives.
Hybrid Home-Work Computing Environment – the resulting networking environment that is created when a user's HC environment (i.e., personal network) is connected to a work computing environment (i.e., organization network) to create a new computing environment that brings with it some of the characteristics and properties of both computing environments.
Internet of Things (IoT) – everyday objects that are embedded with technology to allow that object to be a part of the digital world.
Nudge – changes in the decision-making process that can predictably alter users' behaviors.
Warning – alerts users when a potentially dangerous action has already occurred or when danger is imminent.
Protection Motivation Theory – describes how individuals are motivated to react in self-protective ways towards a perceived threat.
Theory of Planned Behavior – posits that the intention to perform a behavior is the main indicator of a specific behavior.

REFERENCES

Abomhara, M., & Køien, G. M. (2015). Cyber security and the Internet of Things: Vulnerabilities, threats, intruders, and attacks. *Journal of Cyber Security and Mobility, 4*(1), 65–88. https://doi.org/10.13052/jcsm2245-1439.414

American Psychological Association. (n.d.). Health-belief model. In *APA Dictionary of Psychology*. Retrieved May 24, 2022, from https://dictionary.apa.org/health-belief-model

Cain, A. A., Edwards, M. E., & Still, J. D. (2018). An exploratory study of cyber hygiene behaviors and knowledge. *Journal of Information Security and Applications, 42*, 36–45. https://doi.org/10.1016/j.jisa.2018.08.002

Campbell, D. (2019, May 2). *Redline: The Many Human Errors that Brought Down the Boeing 737 Max*. The Verge. Retrieved May 24, 2022, from www.theverge.com/2019/5/2/18518176/boeing-737-max-crash-problems-human-error-mcas-faa

Caraban, A., Karapanos, E., Gonçalves, D., & Campos, P. (2019). 23 ways to nudge: A review of technology mediated nudging in human-computer interaction. In *Proceedings of the 2019 CHI Conference on Human Factors in Computing Systems, 36*, 1–15. https://doi.org/10.1145/3290605.3300733

Clemente, D. (2021). Personal protection: 'Cyber hygiene'. In Paul Cornish (Ed.), *The Oxford Handbook of Cyber Security* (pp. 361–376). Oxford University Press. https://doi.org/10.1093/oxfordhb/9780198800682.001.0001

Detweiler, J. B., Bedell, B. T., Salovey, P., Pronin, E., & Rothman, A. J. (1999). Message framing and sunscreen use: Gain-framed messages motivate beach-goers. *Health Psychology, 18*(2), 189–196. https://doi.org/10.1037/0278-6133.18.2.189

Dodel, M., & Mesch, G. (2017). Cyber-victimization preventive behavior: A health belief model approach. *Computers in Human Behavior, 68*, 359–367. https://doi.org/10.1016/j.chb.2016.11.044

Dupuis, M., Jennings, A., & Renaud, K. (2021). Scaring people is not enough. *Proceedings of the 22nd Annual Conference on Information Technology Education, 22*, 35–40. https://doi.org/10.1145/3450329.3476862

Google Cybersecurity Action Team. (2021). *Threat Horizons: Cloud Threat Intelligence November 2021* [White Paper]. https://services.google.com/fh/files/misc/gcat_threathorizons_full_nov2021.pdf

Howe, A. E., Ray, I., Roberts, M., Urbanska, M., & Byrne, Z. (2012). The psychology of security for the home computer user. *2012 IEEE Symposium on Security and Privacy*. 209–223. https://doi.org/10.1109/SP.2012.23

Kweon, E., Lee, H., Chai, S., & Yoo, K. (2019). The utility of information security training and education on cybersecurity incidents: An empirical evidence. *Information Systems Frontiers*. https://doi.org/10.1007/s10796-019-09977-z

Lexico. (n.d.). Cyber. In *Lexico.com*. Retrieved May 24, 2022, from www.lexico.com/definition/cyber

Li, L., He, W., Xu, L., Ash, I., Anwar, M., & Yuan, X. (2019). Investigating the impact of cybersecurity policy awareness on employees' cybersecurity behavior. *International Journal of Information Management, 45*, 13–24. https://doi.org/10.1016/j.ijinfomgt.2018.10.017

Momenzadeh, B., Dougherty, H., Remmel, M., Myers, S., & Camp, L. J. (2020). Best practices would make things better in the IoT. *IEEE Security & Privacy, 18*(4), 38–47. https://doi.org/10.1109/msec.2020.2987780

Nobles, C. (2019). Establishing human factors programs to mitigate blind spots in cybersecurity. *MWAIS 2019 Proceedings*. 1-6.https://aisel.aisnet.org/mwais2019/22

OWL Labs. (2021). *State of Remote Work 2021*. Retrieved May 24, 2022, from https://owllabs.com/state-of-remote-work/2021

Still, J. D. (2016). Cybersecurity needs you! *Interactions, 23*(3), 54–58. https://doi.org/10.1145/2899383

Still, J. D., Still, M. L., & Grgic, J. (2014). Designing intuitive interactions: Exploring performance and reflection measures. *Interacting with Computers, 27*(3), 271–286. https://doi.org/10.1093/iwc/iwu046

van Bavel, R., Rodríguez-Priego, N., Vila, J., & Briggs, P. (2019). Using protection motivation theory in the design of nudges to improve online security behavior. *International Journal of Human-Computer Studies, 123*, 29–39. https://doi.org/10.1016/j.ijhcs.2018.11.003

Vishwanath, A., Neo, L. S., Goh, P., Lee, S., Khader, M., Ong, G., & Chin, J. (2020). Cyber hygiene: The concept, its measure, and its initial tests. *Decision Support Systems, 128*, 113160. https://doi.org/10.1016/j.dss.2019.113160

Vishwanath, A. (2021). Stop telling people to take those cyber hygiene multivitamins. In M. Khader, G. Ong, & C. Misir (Eds.), *Prepared for Evolving Threats: The Role of Behavioural Sciences in Law Enforcement and Public Safety*, 225–240. World Scientific. https://doi.org/10.1142/9789811219740_0014

Westcott, K. (2019, December 4). *Connectivity and Mobile Trends Survey*. Deloitte United States. Retrieved May 24, 2022, from www2.deloitte.com/us/en/pages/about-deloitte/articles/press-releases/deloitte-launches-connectivity-mobile-trends-survey.html

Xiong, A., Proctor, R. W., Yang, W., & Li, N. (2018). Embedding training within warnings improves skills of identifying phishing webpages. *Human Factors: The Journal of the Human Factors and Ergonomics Society*, *61*(4), 577–595. https://doi.org/10.1177/0018720818810942

Yang, W., Xiong, A., Chen, J., Proctor, R. W., & Li, N. (2017). Use of phishing training to improve security warning compliance. *Proceedings of the Hot Topics in Science of Security: Symposium and Bootcamp on–HoTSoS*, *4*, 52–61. https://doi.org/10.1145/3055305.3055310

9 Will a Cybersecurity Mindset Shift, Build, and Sustain a Cybersecurity Pipeline?

Augustine Orgah
Department of Physics and Computer Science,
Xavier University of Louisiana
New Orleans, LA, USA

INTRODUCTION

It is no secret that there's a shortage of "skilled" cybersecurity professionals needed to fill the large number of openings. As of April 5, 2022, basic searches for cybersecurity jobs on Ziprecruiter.com and cyberseek.com yielded 22,435 and 597,767 job openings in the US, respectively [1][2].

In 2021, 82% of breaches involved social engineering, which involves human elements: social attacks, errors, data mishandling, compromised credentials, and poor security postures to name a few.

A lot of emphasis is placed on the technical ability and competence of a cybersecurity professional with regard to ensuring confidentiality, integrity, availability, and accessibility of data, systems, and services. A lot more will need to be done on the behavioral aspect of cybersecurity [9] to slow down if not thwart the attacks caused by social and behavioral factors.

This research aims to determine if a mindset shift is needed to better prepare cybersecurity professionals to deal with the increasing amount of social and behavioral

DOI: 10.1201/9781003415060-12

attacks. Will a mindset shift that places an emphasis on behavioral cybersecurity improve recruitment and retention in the cybersecurity field overall?

Cybersecurity professionals were presented with a series of questions to determine their feelings about the impact of behavioral cybersecurity: social, behavioral, and cognitive aspects of cybersecurity [9]. They had experience varying from one month to over 20 years.

Undergraduate students, primarily African-American, with no prior knowledge about cybersecurity except an hour of instruction introducing cybersecurity and a list of some tools, partook in a Capture the Flag (CTF) event. They were asked to describe their experience and feelings about the CTF exercise.

In this paper, I will report on both findings, that of the survey and observations about the impact behavioral aspects of cybersecurity in the field and from the undergraduate subjects.

THE SHIFT

The educational pipeline does not seem to be meeting that demand for professionals or the pervasive social engineering attacks going by the rate of attacks and breaches reported in the past year. This paper attempts to analyze the impact that behavioral cybersecurity may yield with current cybersecurity professionals and those attempting to get into the field.

Most breaches in 2021 were caused by attacks that utilized compromised credentials, privilege abuse, and data mishandling. Lots of these attacks utilize well-known techniques yet the malicious actors seem to have the advantage in the race to keep services and infrastructure safe, secure, and available [3][4][6]. The LAPSUS$ extortion group breached its victims primarily using social engineering. They misled, bribed, and impersonated employees of organizations to gain access and implement their malicious agenda. Microsoft, NVIDIA, Samsung, and Okta are among the companies that fell victim to their attacks [8].

The average cost of a security breach rose to the highest value in 17 years of reporting according to IBM from $3.86m to $4.24m in 2021 [3]. Financial gains are the leading motive for 100% of breaches in 2021 according to the Verizon data breach investigations [4]. A data breach cost Equifax up to $425m in settlement fees. The Colonial Pipeline breach initially cost $4.4m in ransom but about 50% of it has been recovered since June 2021 [5].

To aid this analysis, two subject groups were surveyed at different times and using different strategies. The first group comprised of African-American undergraduate students who were thrust into a CTF event without any prior experience. The students provided their reactions using short descriptions of their experience and exposure via email response. The second group consisted of cybersecurity professionals with approximately seven years of experience on average. They were provided with a survey of 12 questions ranging from the amount of experience, education, and the current culture of cybersecurity.

RESEARCH METHODOLOGY

The contents in Table 9.1 constitute the survey questions for the cybersecurity professionals. The survey was provided to the professionals via a URL. The survey was developed with behavioral and general cybersecurity questions as the focus.

The undergraduate students were surveyed via email. They were asked to send an email describing their experience participating in the CTF event and what they learned.

The survey questions were provided to 18 cybersecurity professionals in varying stages of their careers from those having only a month worth of experience to those having over 25 years' experience in cybersecurity. On average, the professionals had approximately seven years of experience. They were asked to provide input using short answers, dropdowns, and select categories. Table 9.2 is a snapshot of the results of the survey.

The second group of this study were undergraduate students at Xavier University of Louisiana in New Orleans. Mostly sophomores, the eight undergraduate students participated in a CTF exercise. The class of 12 students received an hour of preparation and introduction to some cybersecurity tools, and the purpose of a CTF exercise. However, the students were queried by email and only eight of the 12 responded. Traditionally, CTF exercises are staged as a competition and participants compete individually. In the case of our subjects, they were placed in random groups and encouraged to not only work together to find flags but to ask for clues as necessary.

TABLE 9.1
Survey Questions–Cybersecurity Professionals

#	Question
1	**How long have you been a cybersecurity professional?**
2	**Do you consider yourself a cybersecurity expert and/or feel knowledgeable?**
3	**Did you have any formal cybersecurity training before becoming a professional?**
4	**In your opinion, what do you think keeps folks away from the cyber/security field?**
5	**Did/Do you feel intimidated about becoming a cybersecurity professional?**
6	**Is a cybersecurity mindset shift needed to build and sustain a cybersecurity pipeline?**
7	**Do you believe that teaching kids about cybersecurity early will better prepare them for jobs and keep them engaged in the field/industry?**
8	**In your opinion, should cybersecurity education focus on behavioral cybersecurity–cognitive, behavioral, and social aspects? Why?**
9	**Should cybersecurity education be modified to train and attract talent?**
10	**Do you believe that if kids are taught about security early, albeit behaviorally, they will make better cyber professionals?**
11	**Do you believe behavioral cybersecurity education can have a positive impact on recruitment and retention?**
12	**In your opinion, what sort of reputation does the cybersecurity field have?**

TABLE 9.2
Survey Responses–Cybersecurity Professionals

#	Question	Responses		
1	**How long have you been a cybersecurity professional?**	++Average: 7 years		
2	**Do you consider yourself a cybersecurity expert and/or feel knowledgeable?**	**		
3	**Did you have any formal cybersecurity training before becoming a professional?**	Yes: 61%	No: 39%	
4	**In your opinion, what do you think keeps folks away from the cyber/security field?**	+Short answers		
5	**Did/Do you feel intimidated about becoming a cybersecurity professional?**	Yes: 17%	No: 50%	Somewhat: 33%
6	**Is a cybersecurity mindset shift needed to build and sustain a cybersecurity pipeline?**	Yes: 89%	Other Idea: 11%	
7	**Do you believe that teaching kids about cybersecurity early will better prepare them for jobs and keep them engaged in the field/industry?**	Yes: 89%	Other Idea: 11%	
8	**In your opinion, should cybersecurity education focus on behavioral cybersecurity–cognitive, behavioral, and social aspects? Why?**	Yes: 72%	No: 11%	Somewhat: 17%
9	**Should cybersecurity education be modified to train and attract talent?**	Yes: 72%	No: 11%	Somewhat: 17%
10	**Do you believe that if kids are taught about security early, albeit behaviorally, they will make better cyber professionals?**	Strongly Agree: 56%	Agree: 44%	
11	**Do you believe behavioral cybersecurity education can have a positive impact on recruitment and retention?**	Strongly Agree: 33%	Agree: 56%	Neutral: 11%
12	**In your opinion, what sort of reputation does the cybersecurity field have?**	+Short answers		

+ Short answers excluded for brevity
++ Professional Experience (Years) **15–25**: 39%; **4**: 6%; **1–2**: 28%; **1–6 months**: 28%
** Use of Likert scale (1 to 5) Trainee/Minimum knowledge to Expert **1:** 11%; **2:** 39%; **3:** 22%; **4:** 17%; **5:** 11%

ANALYSIS OF RESULTS

There are several conclusions that can be drawn from the results in Table 9.2 involving the cybersecurity professionals. The level of expertise or formal cybersecurity education did not affect belief about the mindset shift believed to be needed to improve cybersecurity. It also did not deny the importance of behavioral cybersecurity and the need for it to be an area for continuous focus. It is interesting that the level of intimidation for/of the field is evenly split. Half of the respondents have no fear, while another 17% of professionals feel intimidated by the field, and 33% feel somewhat intimidated.

The majority of the professionals, 89%, believe that a mindset shift is needed to build and sustain a cybersecurity pipeline. Similarly, the professionals believe that teaching children about cybersecurity early will better prepare them for jobs and retain them in the field.

With regards to behavioral cybersecurity, cognitive, behavioral, and social aspects of cybersecurity, 72% of the respondents believe that cybersecurity education should focus on behavioral cybersecurity. A further 17% were neutral about this focus while 11% do not believe that cybersecurity education should focus on behavioral cybersecurity. The results show that all the respondents support children being taught about behavioral cybersecurity early, which will make them better cybersecurity professionals. The majority of the professionals, 89%, agree that education in behavioral cybersecurity will impact recruitment and retention positively.

There are several conclusions that can be drawn from the undergraduate respondents. They were asked to describe their experience participating in a CTF event after receiving an hour of instruction about the CTF, cybersecurity, and some tools. The students would not have been able to solve the challenges and recover flags without help. After all, they had very minimal exposure to cybersecurity and no experience with CTFs. Despite their limited knowledge, words like "fun", "teamwork", "problem solving", and "career path" were used to describe their experience. Most of them described feeling excited that despite their inexperience they were able to solve some challenges and retrieve flags. They described the experience of learning how to utilize a new tool quickly via hands-on experience and utilizing Google searches. They expressed discovering a potential career path in cybersecurity and that paths exist via the area of computer science.

DISCUSSION–ANALYSIS OF STUDY OPINIONS

There are several conclusions that can be drawn from the results in Table 9.2 involving the cybersecurity professionals. The level of expertise or formal cybersecurity education did not affect the belief that a mindset shift is needed to improve cybersecurity. It also did not deny the importance of behavioral cybersecurity and the need for it to be an area for continuous focus.

The level of intimidation for/of the field is evenly split. Half of the respondents have no fear, while another 17% of the professionals feel intimidated by the field, and 33% feel somewhat intimidated. The respondents either had a born-to-do-it mentality or felt imposter syndrome, not good enough, or were simply intimidated. This was true across all the years of experience.

The majority of the respondents, 72%, believe that cybersecurity education should focus on behavioral cybersecurity. Many of the respondents refer to behavioral aspects as the "people factor." If malicious actors are utilizing psychology, so should security professionals learn about and utilize psychological means to defend and secure. The fact that over 80% of breaches involve some form of social engineering, human social factors means that behavioral cybersecurity cannot be ignored. A professional with adequate behavioral knowledge will better understand the adversary, but importantly be a professional with comprehensive knowledge. It will be accepted that behavioral cybersecurity is a vital portion of the cybersecurity puzzle that can no longer be ignored and should be included in the general cybersecurity education pipeline.

The results show that all the respondents support children being taught about behavioral cybersecurity early, which will make them better cybersecurity professionals. The majority of the professionals, 89%, agree that education in behavioral cybersecurity will impact recruitment and retention positively. The majority of respondents support the mindset shift needed to build and sustain a cybersecurity pipeline. They believe that the marketing and relevance of cybersecurity as a field should improve to show just how broad careers within this field are and that they are available. Recruitment should not be talent-driven alone, but finding individuals with drive, persistence, and motivation. This also ties into making the field an easier entry point for individuals looking to change careers into cybersecurity. Most individuals with behavioral skills will be well suited. Current post-secondary education will need to adapt to keep up with the changes and find creative ways for interdisciplinary education to form a foundation of cybersecurity. It will incorporate behavioral aspects into education, certification, bootcamps, recruitment, and conversions.

Gender was not considered to gauge if females feel different than males about certain survey questions. This could be interesting to evaluate if gender also plays a significant role toward education, recruitment, and retention. It is noted that technology fields are usually male dominated, so perhaps there is an opportunity to examine responses per gender.

The undergraduate students expressed discovering cybersecurity and a potential new career path in computer science, but a wider range of students will need to be surveyed in a more detailed and directed way, similar to the cybersecurity professionals in Table 9.1. It will be beneficial to gauge if the students would feel positive about the cybersecurity field if they participated as individuals and not in groups. None of the students mentioned being intimidated, which was very positive. Nudges and clues were provided as needed to the students as they tried to find flags. However, it will be interesting to get their impressions without the ability to attain clues or nudges.

It also shows that cybersecurity should be taught earlier and included in computational education programs. This small sample set of students were two years into a computer science degree with little to no cybersecurity exposure formally. This is not uncommon looking at four-year computational programs curricula across the country. This is gradually changing as many programs now have cybersecurity concentrations in both undergraduate and graduate level programs. More needs to be done to improve pre-secondary programs to begin early. Make security familiar, interesting, and a way to solve problems, in order to hone security habits, recruit, and retain people interested in security, thereby sustaining a pipeline for cybersecurity positions. In other words, security becomes part of a lifestyle that hopefully will lead to a fine, well-rounded security population and professionals behaviorally and technically.

CONCLUSIONS

This research aims to determine if a mindset shift is needed to better prepare cybersecurity professionals to deal with the increasing amount of social and behavioral attacks. Will a mindset shift that places an emphasis on behavioral cybersecurity improve recruitment and retention in the cybersecurity field overall? There is evidence that emphasis be placed on behavioral cybersecurity as a foundation or alongside the traditional learning paths for analysts, engineers, or other technical positions. This will produce a well-rounded professional with the ability to better deal with behavioral, traditional, and hybrid attacks.

Responses from the professionals and undergraduate students suggest that security should not only be reactive but proactive and a mindset taught and practiced from grade school onwards. It also suggests that we modify the cybersecurity mindset as an approach to collective security, building, and maintaining the pipeline to meet the security needs.

This research has shed some light on what could be done to improve and sustain a pipeline. It highlights the need for a wholistic and inclusive mindset to cybersecurity education that includes behavioral aspects and an introduction early enough that will continue onwards. It should also be marketed differently to highlight the different career options and skills needed. Perhaps a change in approach is warranted. It is laudable that several organizations such as IBM and NSA [11][12] are trying to meet the cybersecurity shortage by creating cybersecurity training centers to meet this challenge.

ACKNOWLEDGMENTS

First, I thank Dr. Wayne Patterson for the opportunity to be a collaborator for this book. I am very grateful! Thank you to Dr. Andrea Edwards of Xavier University of Louisiana for her contributions to this project and the undergraduate surveys. Thank you to the security professionals who took the time to complete the surveys for this research project. Those who want to remain anonymous and to Andrew Tucker, Armando Bazan, Charles Poston, Chase Theodos, Christopher Laibach, Christopher Magill, Cyrus Lee Gerald Robinson, David Londono, Jessica Willoughby, Joseph K. Jaubert III, Rebecca Futch, Sean Scully.

REFERENCES

[1] ZipRecruiter, 2022. [Online]. www.ziprecruiter.com/Jobs/Federal-Cybersecurity

[2] Cyber Seek, 2022. *Cybersecurity Supply/Demand Heat Map*. [Online]. www.cyberseek.org/heatmap.html

[3] IBM, 2021. *How Much Does a Data Breach Cost?* [Online]. www.ibm.com/security/data-breach
[4] Verizon, 2022. *Data Breach Investigation Report.* [Online]. www.verizon.com/business/resources/reports/dbir/
[5] Federal Trade Commission, 2022. *Equifax Data Breach Settlement.* [Online]. www.ftc.gov/enforcement/refunds/equifax-data-breach-settlement
[6] JumpCloud, 2022. *Top 5 Security Breaches of 2021.* [Online]. https://jumpcloud.com/blog/top-5-security-breaches-of-2021
[7] The Seattle Times, 2021. *Capital One to pay $190M settlement in data breach linked to Seattle woman.* [Online]. www.seattletimes.com/business/capital-one-to-pay-190m-settlement-in-data-breach-linked-to-seattle-woman/
[8] Krebs on Security. 2022. *A Closer Look at the LAPSUS$ Data Extortion Group* https://krebsonsecurity.com/2022/03/a-closer-look-at-the-lapsus-data-extortion-group/
[9] Patterson, Wayne & Winston-Proctor, Cynthia. 2019. *Behavioral Cybersecurity*, CRC Press, Orlando, FL
[10] Patterson, Wayne. et al. 2020. The Impact of Fake News on the African-American Community. In Proceedings of Applied Human Factors and Ergonomics (AHFE) International Conference on Human Factors in Cybersecurity, Springer, Virtual/San Diego, USA
[11] IBM, 2022. *IBM Security Learning Academy.* [Online]. www.securitylearningacademy.com/
[12] NSA, 2022. *NSA Centers of Academic Excellence.* [Online]. www.nsa.gov/Academics/Centers-of-Academic-Excellence/

Section IV

Ethical Behavior

10 Cybersecurity Behavior and Behavioral Interventions

Brandon Sloane
New York University
New York, NY, USA

INTRODUCTION

When we think about cybersecurity, it's easy to focus on the technology. We think about computers and networks. We think about software and vulnerabilities. What we don't always remember is that it's just as—if not more important—to consider the people involved. The people are the ones who make decisions about how to build the systems, how to operate those systems, what constitutes safe behavior, and how to respond to potentially risky situations or attacks. The best technological solutions in the world won't protect a company if the people involved aren't making smart and cybersecurity aware decisions.

THE PLAYERS

There are many different ways to think about the players involved in a cybersecurity ecosystem, but for our discussion we will use some simple categories: Attackers, Defenders, Builders, and Community. This conveniently allows us to use the ABCD acronym for our categories. Even though the sequence of terms is "ADBC", we use "ABCD" as a convenient memory reference.

DOI: 10.1201/9781003415060-14

Attackers refer to those adversarial agents who intend malice or present some form of a threat to a company. This class of individuals are also often referred to as Threat Actors or Adversaries. They may include cyber terrorists, state-sponsored actors, organized cybercriminals, hacktivists, or even insiders. Depending on how you categorize these different subgroups, you will emphasize different characteristics and behavioral attributes.

Defenders refer to those protectorate agents who desire to prevent these malicious acts and preserve the confidentiality, integrity, and availability of the assets of a company. This class of individuals are those whom you would typically consider the operators of the systems that protect a company. They may be engaged in protection, detection, or even recovery activities.

Builders are the group of people who design and implement the various systems, which both operate and protect the business functions of a company. While we could potentially group these people in with the Defenders category, we choose to keep them separate because the sorts of behavior that we want to drive or encourage for Builders tends to be different from that which the Defenders perform.

Finally, **Community** refers to the general population of users, employees, contractors, and any other parties who make use of the assets and systems to actually deliver the products and services of the company. This is almost always the broadest category, the easiest to understand, and arguably, the least prepared to act in a way that promotes the cybersecurity hygiene of a company.

Each of these different categories of players in the cybersecurity ecosystem will have different motivations, abilities, skill levels, and any number of other characteristics which will ultimately drive their observed behaviors. As an example, an Attacker might desire to gain access to a list of credit cards in the hopes of using them for financial gain and we might observe them attempting to guess a password to a financial system. A Defender might be highly skilled at reviewing access logs and we might observe high levels of accuracy in their ability to detect malicious login attempts. A Builder might have deep knowledge of access management security and we might observe applications with built-in account lockouts after too many invalid password attempts. Finally, Community members might have very poor cybersecurity awareness and we might observe them using sticky notes on their monitor to remember their passwords.

CYBERSECURITY BEHAVIOR

Having defined the classes of players within the cybersecurity ecosystem, we now need to have a discussion around what constitutes behavior. We think of **Cybersecurity Behavior** as the actions taken by an individual to either increase or decrease the degree to which an asset or set of assets are vulnerable to one or more threats. Pulling from our earlier player conversation as an example, a Community member who decides to post their password on a sticky note has taken an action that increases the vulnerability of a system to the threat of password compromise or credential theft.

As an information security professional, our goals are to reduce the likelihood of potential risks becoming reality. We want to prevent the threats from being successful; to encourage cybersecurity behaviors that decrease the degree to which our assets are vulnerable to threats and discourage those that might increase the vulnerability levels. This general principle applies whether we are discussing the behavior of Attackers, Defenders, Builders, or members of the Community. For example, if we can discourage Attackers from repeatedly attempting to guess passwords, the likelihood of them gaining access will go down. Similarly, if we can encourage Defenders or Builders to use encrypted communications, the likelihood of those messages being intercepted will go down. In both examples, we are reducing our overall risk posture.

We will call this the Fundamental Law of Cybersecurity Behavior:

Encourage behavior which Decreases risk and
Discourage behavior which Increases risk.

DECISION-MAKING MODELS

In order to understand how to drive desirable or discourage undesirable behavior, it is important to take a step back and make sure we have a good understanding of how behavioral decisions are made. We present the following three-step process, which is a modified version of one originally presented over 50 years ago (Locke 1968) [1] and is close to a similar version presented 30 years ago (Jones 1991) [2] in Figure 10.1.

Leveraging this model, we argue that every decision regarding a behavior which an individual makes must first begin with some form of identification that the decision exists. From there, an individual will arrive at some form of a preliminary intention for how to respond to this decision before ultimately acting upon those intentions.

If you look closely, you'll notice there is an extra circular leading back into the Intention step in the process. This is intentional (pun intended) as the Intention step can be much more dynamic and the results of which can and will change over time. As an individual reflects upon or is influenced by various factors, their decision or intended behavior may change as well. In contrast, the Identification and Action steps do not typically have this dynamism. It doesn't necessarily make sense to identify that a decision needs to be made more than once, nor does it make sense that one can perform the same action multiple times. While we could certainly expand our model to allow for those cases, it's easier to argue that those represent new iterations of the decision-making process.

Expanding on the model further, we need to introduce a few additional elements. We first note that for every step in the process, there are generally some inputs and outputs. Further, there are what we call mitigating factors which can affect both how

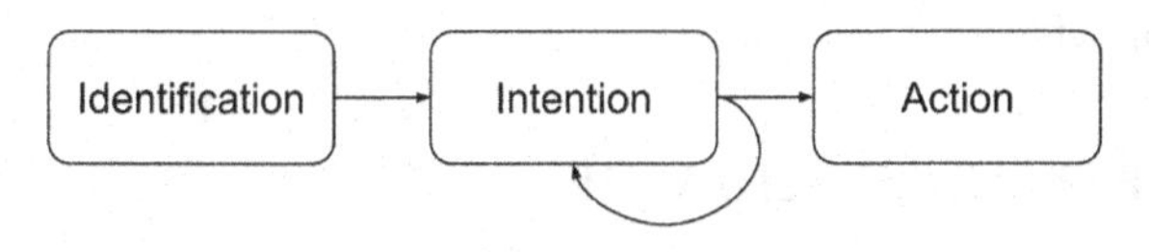

FIGURE 10.1 Behavioral Decision Model.

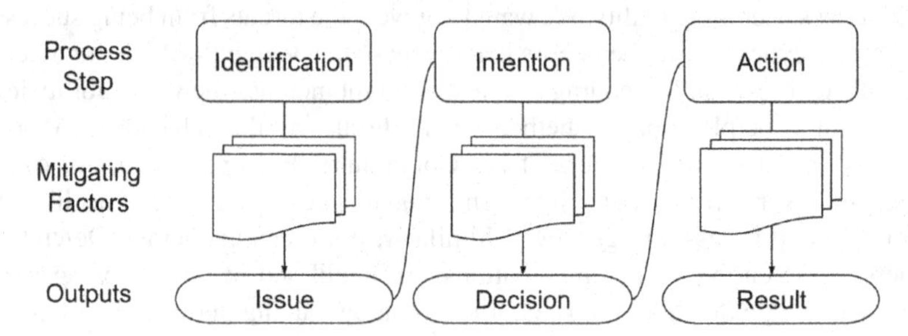

FIGURE 10.2 Behavioral Decision Model Details.

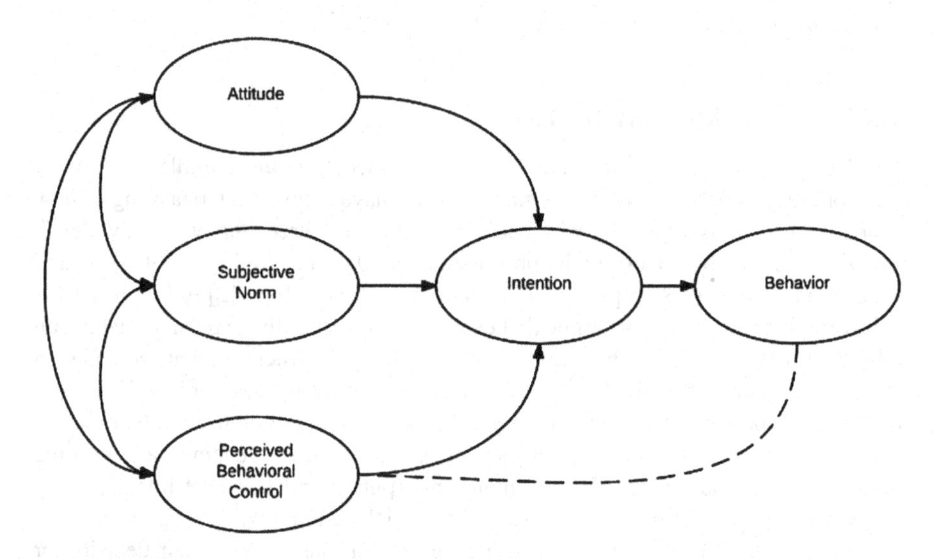

FIGURE 10.3 Theory of Planned Behavior.

and with what result a step in the decision model process might occur. We can graphically indicate these ideas as follows:

Looking at Figure 10.2, a natural follow-up question is: What are these mitigating factors? To answer that, we turn to the body of research into human behavior and various models that have been developed over the years. While there are too many different behavioral models to enumerate here, we will highlight a few specific studies that have identified some key factors. We will then present a non-exhaustive, but fairly comprehensive, listing of factors that are worth considering in a broader meta-model of behavioral studies.

THEORY OF PLANNED BEHAVIOR

The first behavioral model we will discuss is titled a Theory of Planned Behavior (Ajzen 1985) [3] demonstrated in Figure 10.3.

This theory attempted to explain Intention using a combination of three distinct mitigating factors. Ajzen's research measured the extent by which these factors influenced Intention and ultimately Behavior. **Attitude** is an individual's way of thinking or feeling about something. **Subjective Norms** are perceived social pressure to engage or not engage in a behavior. This is typically determined by the total set of normative beliefs weighted by the corresponding motivation to comply. Finally, **Perceived Behavioral Control** is belief about the capability of exerting influence on both internal states and behaviors and/or external environments.

This is a good model to begin with as it is fairly straight-forward to understand all three of these mitigating factors. Further, these factors are all potential levers which we can leverage in our later discussion of driving desired cybersecurity behaviors. For example, Attitude as a factor that drives Community behavior might include the relative acceptance or rejection of something like an information security policy. If the Community thinks of the policy document as a hindrance, something that inhibits their activities, then their Attitude would be generally more negative and we might see less of an Intention to adhere to that policy.

PROTECTION MOTIVATION THEORY

The second behavioral model we will look at is titled Protection Motivation Theory (Rogers 1975) [4] demonstrated in Figure 10.4.

In the interest of brevity, we'll restrict ourselves to just looking at the middle column in Figure 10.4 where Rogers lays out several factors that he was able to demonstrate,

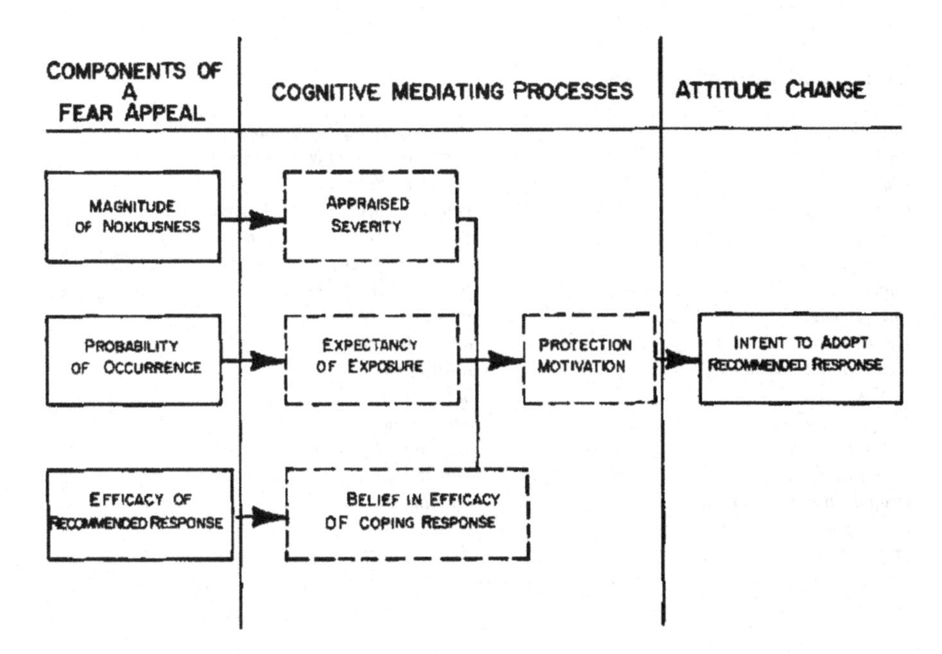

FIGURE 10.4 Protection Motivation Theory.

which contribute towards the behavioral intention decision. These factors include Severity, Exposure, and Belief in Efficacy. **Appraised Severity** will sometimes be referred to as Perceived Severity and deals with an individual's perceptions about the seriousness of a problem that might result from a particular behavior (or lack of behavior). **Expectancy of Exposure** in a cybersecurity context we might more commonly refer to as Perceived Vulnerability and pertains to an individual's perceptions about weakness to adversarial exploitation or action. Finally, **Belief in Efficacy** speaks to an individual's perceptions about the relative impact of their choices and/or actions on one or more desired outcomes.

Severity and Exposure are particularly important factors to consider in the context of a cybersecurity discussion as they directly translate into common risk measurement concepts such as Impact/Magnitude or Likelihood/Frequency. While we won't launch into a full conversation of risk assessment, it suffices to note that Rogers was able to show through his research (Protection Motivation Theory) that individual perceptions of the relative risk (Severity and Exposure) for a particular situation or behavioral opportunity will influence their intentions and ultimate response.

OTHER MITIGATING FACTORS

Before moving on from the topic of decision-making models, we present a list of some other factors that have been documented and researched over the years:

TABLE 10.1
Example Mitigating Factors

Attitude	*Perceived Benefit of Compliance*
Awareness	*Perceived Cost of Non-compliance*
Capability	*Perceived Fairness*
Climate	*Perceived Justice of Punishment*
Coercion	*Perceived Risk of Detection*
Coping Appraisal	*Perceived Severity*
Education	*Perceived Vulnerability*
Enablement	*Persuasion*
Environment	*Response Efficacy*
Facilitating Conditions	*Restriction*
Habits	*Satisfaction*
Incentivization	*Self-Efficacy*
Normative Beliefs	*Superior's Influence*
Opportunity	*Threat Appraisal*
Organizational Commitment	*Training*
Peer Influence	*Visibility*

While this is a non-exhaustive list, there are certainly a number of factors here that have been shown over the years to have strong correlation with influencing behavioral intentions and ultimately behavioral actions in a variety of situations. Some of these should be recognizable as directly relevant to influencing and driving desired cybersecurity behavior. We will get into these a bit more in the next section.

As a final note about decision-making models and mitigating factors, it is important to remember that each isolated behavioral decision-making process will be unique. Different players might have different motivations and perspectives and might therefore place a different weighting on these identified factors. Further, each time a behavioral decision is approached, the individual making the decision comes into that situation with a different context and set of factors, which again might cause different outcomes. All of this is to say that situations and individuals are unique and how an individual responds to a situation will vary greatly.

BEHAVIORAL INTERVENTIONS

At this point in our discussion, we have identified classes of players (Attackers, Builders, Community, Defenders), established a general understanding of what constitutes Cybersecurity Behavior, defined a high-level decision-making model, and finally identified potential mitigating factors that influence how players make decisions. From here, all that remains is to have a deeper conversation about what information security professionals can do to modify or influence mitigating factors in support of a desired behavioral outcome. We do this through **Behavioral Interventions**.

A good first example of a cybersecurity behavioral intervention is the typical training modules that a company will have employees work through on an annual basis. These modules will often include educational components around why information security is important and guidance around what constitutes good security practice. Recalling our discussion on mitigating factors, it's plain that these training modules are an attempt by companies to influence factors such as awareness, education, and perceived severity. By requiring employees to undergo training on a regular basis, a company can ensure that when faced with cybersecurity behavioral decisions, they are better equipped to identify and ultimately act in the best interests of the company.

There have been a few attempts over the years to develop a framework for organizing different intervention types. There are two in particular that we will discuss here as we realize that having a structured way of thinking about interventions will aid practitioners in developing their intervention strategies.

MINDSPACE

The first behavioral intervention framework we discuss uses the mnemonic MINDSPACE (Dolan et al. 2012) [5] demonstrated in Figure 10.5.

MINDSPACE cue	Behaviour
Messenger	We are heavily influenced by who communicates information to us
Incentives	Our responses to incentives are shaped by predictable mental shortcuts such as strongly avoiding losses
Norms	We are strongly influenced by what others do
Defaults	We 'go with the flow' of pre-set options
Salience	Our attention is drawn to what is novel and seems relevant to us
Priming	Our acts are often influenced by sub-conscious cues
Affect	Our emotional associations can powerfully shape our actions
Commitments	We seek to be consistent with our public promises, and reciprocate acts
Ego	We act in ways that make us feel better about ourselves

FIGURE 10.5 MINDSPACE.

Developed by the UK's Institute of Government, MINDSPACE describes a set of nine different biases or effects that have a strong impact on behavior. Messenger refers to how using different people or delivery methods can impact how the contents of that message are received and processed. Incentives is a category of interventions that encompasses both rewards and punishments. This list is a mix of strategies, attributes, mechanisms, and modes of delivery, but they all share one important characteristic: they can all be used to influence the behavioral decision making.

BEHAVIOR CHANGE WHEEL

The second framework we want to present is the Behavior Change Wheel (Michie et al. 2011) [6] demonstrated in Figure 10.6.

This framework is interesting and useful because it takes the concept of interventions and extends it further by also introducing categories of policies that can be utilized to apply different types of interventions. The Behavior Change Wheel is designed as such that you begin by identifying the sources of behavior in terms of Capability, Opportunity, and Motivation. You then identify one or more categories of Interventions that you want to employ to influence those sources of behavior. Finally, you select one or more Policy categories with which to employ your chosen Interventions.

As an example of how to utilize the Behavior Change Wheel, we might decide that our Community clicks on too many phishing emails and we want to raise their Capability to detect a malicious email from a legitimate one. Having identified Capability as the source of behavior, we might then determine that we want to both Educate our employees on the dangers of clicking on malicious links and also Train them on how to detect a phish. This ultimately leads us to both define a company policy, which requires annual phishing training, and also to launch an internal marketing campaign about the dangers of phishing.

CONCLUSION

In this chapter, we have talked about the who (Players: Attackers, Builders, Community, Defenders), the what (Cybersecurity Behavior), the why (Decision-Making Model), and the how (Behavioral Interventions) of encouraging or discouraging behavioral outcomes. These are the building blocks necessary for companies to make people an active part of their cybersecurity program. Through active identification of desired

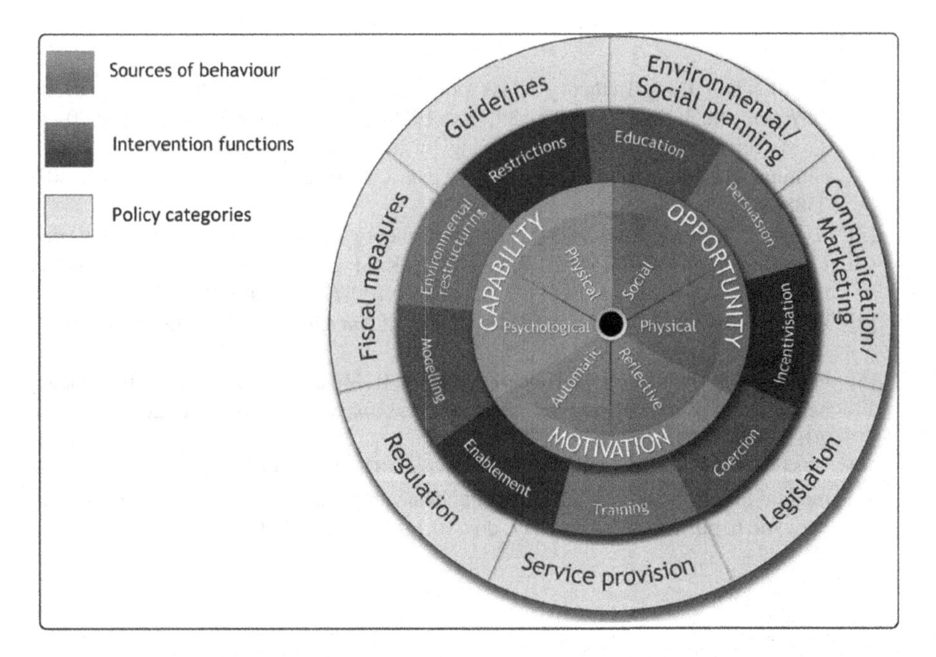

FIGURE 10.6 Behavior Change Wheel.

behaviors and proactive behavioral interventions targeted at distinct populations, companies can moderate people risk and potentially even turn them from a liability into an asset.

QUESTIONS

1. Provide an example of a behavioral decision that an Attacker might make, which could result in a decrease to the risk faced by a company. What sorts of Behavioral Interventions might a company take to encourage the Attacker to make this decision?
2. Under what circumstances might a Behavioral Intervention backfire or have unintended consequences? Provide an example of a real-world situation where a Behavioral Intervention didn't go as planned.
3. With regards to Incentives, do you think that providing financial incentives for desired cybersecurity behavior is a good idea or a bad one? What other kinds of Incentives can you come up with which might be more effective and for what specific populations?
4. Provide an example of a Behavioral Intervention that might work for one population but wouldn't work for another. Similarly, provide an example of a Behavioral Intervention that might work the first time but would be unlikely to succeed under multiple repetitions.

REFERENCES

[1] E. A. Locke, "Toward a theory of task motivation and incentives," *Organ. Behav. Hum. Perform.*, vol. 3, no. 2, pp. 157–189, May 1968, doi: 10.1016/0030-5073(68)90004-4.

[2] T. M. Jones, "Ethical decision making by individuals in organizations: An issue-contingent model," *Acad. Manage. Rev.*, vol. 16, no. 2, p. 366, 1991, doi: 10.2307/258867.

[3] I. Ajzen, "From intentions to actions: A theory of planned behavior," in *Action Control: From Cognition to Behavior*, J. Kuhl and J. Beckmann, Eds. Berlin, Heidelberg: Springer, pp. 11–39, 1985, doi: 10.1007/978-3-642-69746-3_2.

[4] R. W. Rogers, "A protection motivation theory of fear appeals and attitude," *J. Psychol.*, vol. 91, no. 1, pp. 93–114, 1975.

[5] P. Dolan, M. Hallsworth, D. Halpern, D. King, R. Metcalfe, and I. Vlaev, "Influencing behaviour: The mindspace way," *J. Econ. Psychol.*, vol. 33, no. 1, pp. 264–277, Feb. 2012, doi: 10.1016/j.joep.2011.10.009.

[6] S. Michie, M. M. van Stralen, and R. West, "The behaviour change wheel: A new method for characterising and designing behaviour change interventions," *Implement. Sci.*, vol. 6, no. 1, p. 42, Apr. 2011, doi: 10.1186/1748-5908-6-42.

Section V

Differences in Languages in Cyberattacks

11 Using Language Differences to Detect Cyberattacks

Ukrainian and Russian

Wayne Patterson
Patterson and Associates
Washington, DC, USA

Previous studies have shown that detecting translations from one language to another can be an effective tool in determining the source of cyberattacks. This has been demonstrated in research by Blackstone and Patterson (2022) [3], Florea and Patterson (2021) [4], and Patterson, Murray and Fleming (2020) [5].

ASSESSING THE SKILLS OF CYBER ATTACKERS

As a preliminary thought on the development of this technique, it is important to understand that different categories of cyber attackers have differing levels of attack skills. For example, if the cyberattack is being launched by a national government or an attack organization with vast resources, many avenues are open in terms of the use of resources to launch attacks on promising targets.

On the other hand, the resources of a national government or a large criminal organization are unlikely to be wasted by the organization in question if the chances for the reward are far less than the investment required in order to mount the attack.

By the same token, the ordinary defender might have a concern that their personal accounts are likely to be compromised. But a national government or a large criminal

DOI: 10.1201/9781003415060-16

organization also needs to consider the same principle. If the ordinary user does not present an inviting target—for example, by possessing a vast sum of money that can be unlocked by being hacked—then the ordinary user has less to be concerned about if the potential cyber attacker determines that the potential reward itself not worth the effort.

In other words, the scenario for the potential interaction between the cyber attacker and the cyber defender can be modeled by a game theory approach. To take the simplest version of such an analysis, let us reduce the problem to that of only two potential attackers and two potential defenders (Patterson and Winston-Proctor, 2018) [6].

The attackers and defenders are described in the following two-by-two matrix:

		Cyberattacker investment	
		Weak attack	*Strong attack*
Cyberdefender investment	*Weak defense*	$0	-$1000
	Strong defense	+$1000	$0

In this game, if both attacker and defender invest the same amount (say, $1,000) in the attack or defense, the defender will defeat the attack and protect his or her amount; if the attacker mounts a strong attack and the defender a weak defense, the attacker will steal the value in question (by convention, the column player's winnings are represented by a negative number).

Having analyzed this game theory model, it makes it reasonable to assume that the ordinary user, whose economic value might be described in the thousands of dollars in monetary units, is unlikely to be attacked by a very sophisticated attacker who will not want to spend more in the attack than in receiving a potential reward. Consequently, for the ordinary user, it is more important to be prepared to mount a defense against the more elementary type of attacker, which we commonly call "script kiddies".

With this analysis, we can posit several conditions on the potential criminal organization or nongovernmental cyber attacker. The first of these, which we will analyze in more detail in this chapter, is that both the attacker and defender will have to address the situation where the attack conversation must use at least in part some human language, assuming the attacker does not have the resources for extensive and expensive language translation. This then imposes the restriction that the attacker must be able to gain some information about the human languages understood by the victim.

With this assumption, we can then analyze how the attacker can go about formulating an attack dialogue that will be understood in the language of the attacked person, and yet not raise suspicion that the attacker actually has little knowledge of the victim's natural language.

There are many tools to provide automated language translation, of which a number are very sophisticated but are limited in their cost and accessibility.

USING GOOGLE TRANSLATE

For this reason, it is assumed that the best tool available for an attacker to convey an attack scenario in the language of the victim is the software known as Google Translate (GT).

GT is fully accessible to almost all computer users and can translate text between any of 103 different human languages, and it is free to the user.

Thus, it is reasonable to assume that for a cyber attacker with limited resources needing to formulate an attack script in a different human language, will be drawn to the use of GT to bring this about.

In our previous work, we began with the development of two software tools, which, when combined with GT, yields a metric in terms of measuring the efficiency of language translation with GT. This metric involves first a GT translation from one language to another, then back to the first—we call the "ABA translation"; the measure of the difference in the two versions of "A" is computed using a variation of the well-known Levenshtein Distance (Navarro, 2001) [7] that we call Modified Levenshtein Distance (MLD).

We developed a test bed of 10 quotes from English literature and 10 from more colloquial English. The reason for these two sources is to test the difference in the quality of GT between the more formal quotations (as drawn from Bartlett's) (Bartlett, 2012) [2] and quotations from popular dialogue, in this case well-known quotes from films as compiled by the American Film Institute (AFI, 2022) [1]. Here is a sample of some of the translations used in the development of this test bed:

A) ERE English-Romanian-English:

Romanian: Dintre toate articulațiile de gin din toate orașele din întreaga lume, ea intră în a mea.

Of all the gin joints in / all the towns / in all / the world, she
Of all the gin joints in / every city /around / the world, she
/ store 9 ↑ store 6 ↑

/ walks into / mine.
/ enters / mine.
/ store 9 ↑

Thus MLD = 9 + 6 + 9 = 24.

==

B) EGE English-Greek-English:

Greek: Ένα ψέμα φτάνειστα μισάτηςγης πριν η αλήθεια έχειτηνευκαιρία να φορέσειτο παντελόνιτης.

A lie / gets / half / way / around the earth before the truth
A lie /reaches / half / / around the earth before the truth
/ store7 ↑ / store 3 ↑

has a chance to / get its/ pants on.
has a chance to / put on/ her pants.
/ store 6 ↑ store 8↑

Thus MLD = 7 + 3 + 6 + 8 = 24.

===================================

C) ERE English-Russian-English, and EUE English-Ukrainian-English

Russian: Думаете ли вы, что можете или нет, но обычно вы правы.

Whether you think /that /you can, or /that you can't, / you are usually right.
Whether you think / /you can or /not, /you are usually right.
/ store 4 ↑ / store 11 ↑

Thus MLD = 4 + 11 = 15.

Ukrainian: Незалежно від того, чи вважаєте ви, що можете, або що ви не можете, ви, як правило, маєте рацію.

Whether you think / that /you can, or that you can't, you /a/re usually right.
Whether you think / /you can or / /can't, you'/ / re usually right.
12
/ store 4 ↑ / store 7 ↑ / store 1 ↑

Thus MLD = 4 + 7 + 1 = 12.

===================================

D) ERE English-Russian-English, and EUE English-Ukrainian-English

Russian: Скажите им, чтобы они вышли с тем, что у них есть, и возьмите только одно для ловкача.

Tell /' /em to /go /out /there /with /all /they /got /and /win just /one for the /Gipper.

Tell / th/ em to /come /out / /with /what /they /have /and /only take /one for the /dodger.

/ store 2 ↑ / store 4 ↑ / store 5 ↑ / store 4 ↑ / store 4 ↑ / store 8 ↑ / store 6 ↑

Thus MLD = 2 + 4 + 5 + 4 + 4 + 8 + 6 = 33.

Ukrainian: Скажіть їм, щоб піти туди з усім, що вони отримали, і виграти тільки один для Gipper.

Tell /‘ /em to go out there with /all /they / /got and win just one for /the/Gipper.
Tell / th/ em to go out there with / everything /they/‘ve /got and win just one for / /Gipper.
/ store 2 ↑ / store 10 ↑ /store 2 ↑/ store 3 ↑

Thus MLD = 2 + 10 + 2 + 3 = 17.

====================================

For a specific test case for this paper, we are going to reflect on the current global controversy involving the invasion of Ukraine by Russia and try to determine some criteria to see if we can detect text that might have been developed in Russia and is masking as Ukrainian; or vice versa, text developed in Ukrainian that has been translated into Russian.

In this case, we also use two datasets for the comparison. First, we indicate our results based on the aforementioned quotes from English sources as used in [4]. In the second case, we develop a specific vocabulary based on quotes involving the subject of “War”. Here are a few of the examples used:

Samples of “ABA” Translation Referencing “War”
These examples are labeled A through E:

A
Dwight D. Eisenhower (President of the United States, 1952–60)

Russian
Я ненавижу войну, как может только солдат, проживший ее, только как тот, кто видел ее жестокость, ее тщетность, ее глупость.

Original English
I hate war as only a soldier who has lived / /it can, only as one who has seen its / brutality /, its futility, its stupidity.

Back to English
I hate war as only a soldier who has lived /through /it can, only as one who has seen its / cruelty/, its futility, its stupidity.
7 +
9 = 16

Ukrainian
Я ненавиджу війну, як може тільки солдат, який пережив її, тільки як той, хто бачив її жорстокість, її марність, її дурість.

Original English
I hate war as only a soldier who /has lived / it can, only as / /one who / has seen / its / brutality/, its futility, its stupidity.

Back to English
I hate war, as only a soldier who / survived / it can, only as / some/one who / saw / its / cruelty /, its futility, its stupidity.
8 + 4 +
7 + 9 = 28
==

B
Napoleon Bonaparte (First Consul, Republic of France, 1799–1804)

Russian
Вы не должны слишком часто сражаться с одним врагом, иначе вы научите его всему своему военному искусству.

Original English
You must not fight / / too often / with one enemy, / or / you will teach him all your / art of war.

Back to English
You must not fight / one enemy / too often, / / otherwise / you will teach him all your / martial arts.
12 + 9
+ 11 = 32

Ukrainian
Ви не повинні битися занадто часто з одним ворогом, інакше ви навчите його всьому своєму військовому мистецтву.

Original English
You / must / not fight too often with one enemy, / or / you will teach him all your / art of war.

Back to English
You / should / not fight too often with one enemy, / otherwise / you will teach him all your / martial arts.
6 + 9 +
11 = 26
==

C

Salvador Dali (Spanish surrealist artist, 1904–89)

Russian

Войны никогда никому не причиняли вреда, кроме тех, кто умирает.

Original English

Wars have never / hurt / any/body / except / the people / who die.

Back to English

Wars have never / harmed / any/one / except /those / who die.

6 + 4 + 9 = 19

Ukrainian

Війни ніколи нікому не шкодили, окрім людей, які гинуть.

Original English

Wars have never / hurt / any/body / except / the / people / who / die.

Back to English

Wars have never / harmed / any/one / except / / people who / are dying.

6 + 4 + 3 + 8

= 21

==

D

Hiram Johnson (Governor of California, 1911–17)

Russian

Первая жертва, когда приходит война, — это правда.

Original English

The first casualty when war comes is / / truth.

Back to English

The first casualty when war comes is / the / truth.

3 = 3

Ukrainian

Першою жертвою, коли настає війна, є правда.

Original English

The first / casualty / when war comes is / / truth.

Back to English

The first / victim /when war comes is / the / truth.

8 + 3 = 11

==

E
Isaac Asimov (science-fiction author)

Russian
На войне гибнут не только живые.

Original English
It is / not only the living / who are killed / in war.

Back to English
/ / Not only the living / die / in war.
4 + 12 = 16

Ukrainian
На війні гинуть не тільки живі.

Original English
It is / not only the living / who are killed / in war.

Back to English
/ / Not only the living / die / in war.
4 + 12 = 16

EXAMINING THE COMPILED TESTS

In most of the previous studies cited, a common test bed for comparisons of writings in one language translated to another, then translated back, is what we have called ABA translation. In this paper, we try to see if a test bed of phrases about warfare might point to different comparisons, since recent warfare involving Russia and the Ukraine might use more often certain terminology related to warfare.

Of course, any choice of phrases or sentences translated between two languages will only yield a small set of data points in the comparison. Nevertheless, what we find in a general sense is that the accuracy of ABA translation substantially varies between the test bed that has been used in several studies cited, whereas the "War" test bed is unique to this study.

Nonetheless, the differences in translation in these two distinct cases is demonstrated by the following table. With the test bed of "War" quotes, in about two-thirds of the overall cases, there are large disparities in the Levenshtein Distance between the Ukrainian case and the Russian case, but the cases in which the EUE translation is closer, then the ERE translations are about equal in number. On the other hand, using the database developed from literary and movie quotes shows the ERE translations have greater errors in 60% of cases, but in only one case do the EUE error translations have greater errors. However, when using the "War" quotes test bed, the incidence of errors of ERE errors to EUE are very close to the same.

	ERE errors > EUE errors by more than 5%	EUE errors > ERE errors by more than 5%
Literary and movie quotes	12 out of 20 cases	1 out of 20 cases
"War" quotes	20 out of 60 cases	21 out of 60 cases

What we can conclude from this analysis is that in the more general set of test cases, "literary and movie", there is a measurable distinction in detecting Russian and Ukrainian translation, but this distinction essentially disappears with the more specialized vocabulary.

REFERENCES

[1] American Film Institute. AFI's 100 Years … 100 Movie Quotes. www.afi.com/afis-100-years-100-movie-quotes

[2] Bartlett. J. *Bartlett's Familiar Quotations.* Little, Brown and Company: New York, NY.

[3] Blackstone, J., Patterson, W. (2022). *Isolating Key Phrases to Identify Ransomware Attackers.* Proceedings of the 13th International Conference on Applied Human Factors and Ergonomics (AHFE 2022), New York, NY.

[4] Florea, D., Patterson, W. (2021). *A Linguistic Analysis Metric in Detecting Ransomware Cyber-Attacks.* www.thesai.org.

[5] Patterson, W., Murray, A., Fleming, L. (February 2020). Distinguishing a Human or Machine Cyberattacker. Proceedings of the 3rd Annual Conference on Intelligent Human Systems Integration, Modena, Italy. pp. 335–340.

[6] Patterson, W., Winston-Proctor, C. (2019) Behavioral Cybersecurity, CRC Press: Boca Raton.

[7] Navarro, G. (2001). A Guided Tour to Approximate String Matching (PDF). ACM Computing Surveys. 33(1), pp. 31–88. CiteSeerX 10.1.1.452.6317. doi:10.1145/375360.375365. S2CID 207551224.

Section VI

Applications of Behavioral Economics to Cybersecurity

12 Using Economic Prospect Theory to Quantify Side Channel Attacks

Jeremy Blackstone
Howard University
Washington, DC, USA

The main method used to prevent exploitation of consumer products, such as cell phones, smart cards and automobiles, is through mathematically secure cryptographic algorithms [1]. However, side channel analysis (SCA) attacks bypass these algorithms by monitoring the effects of the algorithm on a physical platform through power consumption, electromagnetic emanations (EM), timing of operations, acoustic vibrations, or subjecting the device to fault injection [1]. By measuring these aspects of the physical computations, an attacker is able to discover sensitive information, e.g., extracting the secret key from a cryptographic algorithm. I aim to determine the probability an SCA attack will occur based on human perception of risk.

Since SCA attacks are a relatively new security concern, many are not familiar with the potential risks of hardware that is not equipped to mitigate these attacks. For this reason, I plan to estimate the likelihood of adversaries launching SCAs as well as potential targets implementing SCA countermeasures using economic prospect theory [2]. I will quantify the cost and reward for SCA adversaries as well as the cost and risk for SCA targets. Similar work is done by Patterson & Gergely (2020) [2], but this paper does not correlate the probabilities produced to any specific attacks. This research plans to estimate the costs and rewards based on concrete attack scenarios

DOI: 10.1201/9781003415060-18

for SCA specifically. Furthermore, all test subjects were undergraduate university students in the United Arab Emirates taking a cybersecurity course. I intend to have test subjects from a variety of cultural backgrounds and at different levels of expertise in security. With regard to level of expertise, some participants will be students in a non-STEM field, some participants will be freshman computer science undergraduate students, some participants will be upperclassmen computer science students who have completed cybersecurity courses, some participants will be computer science graduate students who have completed cybersecurity courses and some participants will be cybersecurity experts. With regard to cultural background, some participants will be African-American, some participants will be Caucasian-American and some students will be international students from Africa and Asia.

REFERENCES

[1] A. Barenghi, L. Breveglieri, I. Koren, and D. Naccache, "Fault injection attacks on cryptographic devices: Theory, practice, and countermeasures," *Proceedings of the IEEE*, vol. 100, no. 11, pp. 3056–3076, 2012.

[2] W. Patterson, and M. Gergely, "Economic prospect theory applied to cybersecurity," in *International Conference on Applied Human Factors and Ergonomics*, pp. 113–121, Springer, 2020.

13 A Game-Theoretic Approach to Detecting Romance Scams

Ebelechukwu Nwafor
Villanova University
Villanova, PA, USA

INTRODUCTION

As society becomes increasingly connected through the Internet, romance scams also known as dating scams have become a widespread issue. This involves a nefarious entity known as a scammer creating a false identity with the aim of pursuing a romantic relationship with an unsuspecting victim. The main objective of a scammer is to obtain financial gain or to steal the victim's personal information. Traditional means of detecting romance scams heavily rely on human intervention such as online moderation. However, as the number of online dating interactions continues to increase, it has become challenging for humans to keep up with traditional systems. Furthermore, the use of human moderators to detect romance scams can be inconsistent, as policies may come from different geographical regions making it difficult to adhere to a standard policy. It is important to adopt a comprehensive approach to scam detection that incorporates both automated and human moderation techniques. This will help to mitigate the risk of false positives in the automated systems being overlooked by human review. One potential solution is the use of game theory to provide an automated means of detecting romance scams. Game theory is the study of strategic decision making and has applications in a wide range of problems such as economics, conflict resolution, and computing [1]. This approach provides a framework based on the construction of rigorous mathematical models that can be used to optimize the outcome in decision scenarios thereby improving decision making.

DOI: 10.1201/9781003415060-19

In romance scams, game theory can be leveraged to model interactions between a scammer and a victim in order to minimize the payoff of the scammer. We can identify techniques that are commonly used by the scammer, such as constant requests for funds and attempts to obtain personal information. The system can flag such interactions and provide information to human moderators.

Next, we define a game-theoretic technique that can be utilized to model a romance scam known as the minimax regret.

MINIMAX REGRET

This approach involves minimizing the maximum regret of a player. The regret is a measure of the difference between the best value the player can achieve and the actual value in which the player achieves. In this scenario, a player tries to minimize the regret they feel if things do not go as planned. To achieve this, they evaluate every possible scenario by calculating the regret from each scenario. The player then chooses the option with the least amount of regret from all the possible scenarios regardless of the outcome.

Minimax regret can be an effective strategy that a victim can use to minimize the amount of money or information lost in a romance scam scenario. Next, we provide a simple scenario in which minimax regret can be used to model interactions between the scammer and the victim in romance scams:

> *Suppose a victim receives a $100 monetary request from a scammer, the victim has two possible actions—**to send the money to the scammer** or **not to send money to the scammer**. Based on these options, the possible outcomes to the scammer are to **continue the scam** or **not to continue the scam, and not to continue could be (1) concluding there is no reward likely or (2) fear of detection.***

The regret associated with the choice of the victim's outcomes can be calculated as follows:

- The victim sends the money and the scammer continues with the scam, the regret for the victim is the amount of money that was sent.
- The victim sends the money and the scammer ends the scam, the regret for the victim is the amount sent.
- The victim does not send the money, and the scammer continues to scam, the regret for the victim is the potential loss of any additional money that the scammer might have requested.
- The victim does not send the money and the scammer ends the scam, the regret for the victim is zero. In this case, an optimal choice was made.
- The scammer fears discovery, so plans an alternate approach, costing $300.
- The victim takes a precaution in case of a scam, spending $50 for some form of detection software.

This scenario can be presented in Normal Form as follows:

S.v = Sends Money
S.s = Continues Scam

D.v = Does not Send Money
D.s = Discontinues Scam
T.s = Trap is feared by Scammer
T.v = Trap is anticipated by Victim
Players = {Victim, Scammer}
Actions = {S.v, S.s, D.v, D.s, T.s, T.v}

	Scammer			
Victim		S.s	D.s	T.s
	S.v	-100, 100	-100, 0	-100, 300
	D.v	0, 100	0,0	0, 300
	T.v	-50, 100	-50, 0	-50, 300

The Minimax Regret Criterion

To minimize the "maximum regret", first "normalize" the entries.

In the matrix above, we see that the victim's minimax regret as the scenario where the minimal regret was made. In this case this is when the victim does send money to the scammer (S.v) but also the scammer discontinues the scam (D.s).

Each entry is modified by changing each to the absolute value of each entry comment, for example: {-100, 100} → 200. The matrix above becomes:

	Scammer			
Victim		S.s	D.s	T.s
	S.v	200	100	400
	D.v	100	0	300
	T.v	150	50	350

The minimax regret is found by identifying the element that is the minimal value in its column (D.s) and the maximal value in its row (S.v). In this case, this is the element defined by S.v and D.s, with value 100. Thus, the minimax regret value is 100, and it is associated with the actions S.v and D.s.

In conclusion, this paper provides an overview on how game theory can be used to model romance scam scenarios. Using game theory to develop an automated means for detecting romance scams can be an effective way to protect unsuspecting individuals from financial and personal theft. Furthermore, it can improve the overall quality of interactions across dating sites and increase their credibility, making the Internet safer for everyone.

REFERENCE

[1] W. Patterson and Cynthia E. Winston-Proctor, "Behavioral Cybersecurity", CRC Press 2019 (Boca Raton, FL), *Chapter 13: Game Theory pp. 95–110.*

Section VII

New Approaches for Future Research

14 Human-Centered Artificial Intelligence

Threats and Opportunities for Cybersecurity

Gloria Washington
Howard University
Washington, DC, USA

1 INTRODUCTION

What is Human-Centered Artificial Intelligence?

As the rise of smartphones and devices with Internet capabilities increases globally, the opportunities for human-centered artificial intelligence (HCAI) are limitless as these tools seek to improve every aspect of human life. Innovations like smart homes, medically-implanted devices, healthcare assistant technology, emotional AI that tracks and responds to the physiological reactions associated with feelings, and other smart wearable technology; the benefits of these HCAI are without measure. However the harms for HCAI are also boundless. The HCAI field is concerned with using human physical, physiological, behavioral, and/or other characteristics for creating technology for social good, but often these enormous holes for exploiting data used in HCAI are not properly studied, examined, and mitigated. Often, the capitalist gains of potential HCAI technologies are only thought about if harms are tested and played out in the communities and populations that are the most marginalized.

DOI: 10.1201/9781003415060-21

2 AIMS AND METHODS

The goal of this chapter is to explore an increasing area of technology innovation that seeks to improve and understand the human experience. Because the artificial intelligence (AI) arena hopes that machine-learning (ML) techniques will one day help humans perceive, synthesize, and infer information for helping humans make better decisions, this exploration of study may include non-AI approaches that do not utilize ML techniques. This is because many applications that have the goal to improve human's lives do not need to employ sophisticated ML techniques. Examples of such applications may include those that allow people to report safety concerns and crime in their neighborhoods and crowdsource funding for social improvement projects. Although these applications are human-centered, they do not necessarily use AI to create a unique experience for its users. However, these applications allow humans to collect, curate, and cultivate data relevant to their needs. This personally identifiable data can be exploited by criminal elements wishing to sell it for financial gain or use it to steal or create fake identities. The aims of this chapter are to identify such technologies intended for various industries including marketing, criminal justice, Internet search technologies, social media, and other areas, while noting through observational practices how these technologies can be perceived to be human-centered. Finally, the implications of the threats and opportunities for exploitation of the data are discussed.

The observations included in this research on human-centered technologies include those for understanding sentiment in text, audio, and video data, virtual and augmented reality for simulating human experiences, facial expression recognition and face recognition, voice analytics, social networking applications, emotion-enhanced smart cities, and wearable technologies. Human-centered technologies are being created by companies outside of the tech sector. In fact, this research explores technologies from genres including marketing and advertising, law enforcement, medicine, defense, mental health, graphic arts, filmmakers, gaming, and sporting agencies.

This chapter is organized by providing an overview of past, contemporary, and future technologies in HCAI, while examining the data communicated by these tools. We will first start with explaining biometric technologies or technologies that use human body-based measures including images, text, video, audio, and health signal data. We will include healthcare-related biometrics as well as security-related biometrics for identifying and recognizing humans. We will then explore technologies that utilize human behavior-related information for learning about humans. For each of these explorations, we will identify and examine threats to humans and opportunities for future safeguarding tools. Finally, we will land on a discussion of the limitations surrounding this exploration.

3 BIOMETRICS

3.1 Smart Device Biometrics

Biometrics is a vast field affecting both healthcare and security-related industries. Many people may be familiar with the biometric sensors embedded into their smartphones like the fingerprint sensor or face recognition unlocking technology. Smartphone access control biometrics utilize images of fingerprints and faces to develop a unique biometric profile for locking and unlocking mobile devices. This process includes two steps for properly accessing smart devices using biometrics: enrollment and authentication. The enrollment process includes scanning the physical characteristic, e.g. a face or fingerprint, several times to create a biometric template that is stored either on the device or in the cloud. Several images are often needed to create the biometric template because features are gathered and averaged across several images to create the biometric profile. In the authentication process, a user will scan their biometrics, an image is captured, features are processed across the image, and used to compare against the features stored in the biometric template. Access is given, if and only if, the features match according to a certain accuracy threshold or all features match the biometric template. In the case of fingerprint authentication, most often, all features of the fingerprint compared must match the template stored.

Opportunities for exploiting and improperly accessing the biometric template relate to 1) human error, 2) tampering with the biometric sensor, and/or 3) tampering with the network connection used to communicate the image over to the device. Human error can occur when a user of a smartphone erroneously leaves their smartphone unlocked and the biometric template is altered or changed without the knowledge of the smartphone owner. Furthermore, human error can occur when using the biometric sensor improperly. A buildup of dirt, oils, and other obstructions may impact the quality and integrity of the images captured with the sensor. These images, if collected with a dirty or faulty sensor, can lead to an inaccurate biometric template that does not properly contain the features necessary for matching against future attempts to unlock the phone.

The biometric template can be changed or altered if the sensor used for enrollment purposes is exploited so that data gathered is corrupted or damaged during collection. Cybersecurity professionals stress that using the biometric sensors available on smartphones protects data from unauthorized use. However, smartphone users often don't enable them. Unauthorized users may intentionally damage the biometric sensors for future use so that the smartphone data could be accessible without this added protection. Data gathered from the biometric sensor may be unusable, inaccurate, and unreliable for creating a biometric template. In this case, the owner of the smartphone or smart device may give up during enrollment and not use it to protect their phone.

The biometric template can be exploited and accessed through the network connection that connects the data with the Internet, the network stored on the device, or through a USB connection cord. Using the Internet, unauthorized users can access the biometric template while a user is enrolling or through the authentication process. During enrollment, the user creates a biometric profile; but sometimes applications installed on the device give access to all data collected and communicated across the phone. In this case, an unauthorized user of the smartphone may access, change, corrupt, and/or delete the biometric template either during creation or after it has been created. Many cybersecurity professionals indicate that users must read and understand applications and data access given to these applications. Unfortunately, many smartphone users do not read application disclosures and may unknowingly download malware that collects and sees all data communicated through a device. Additionally, unauthorized access to the biometric template may occur through applications stored on the device. However, the biometric template can be exploited through connection with the Internet. Smartphone and device owners may unknowingly connect to insecure and open WiFi connections to save on mobile data usage. However, many times these open networks are created so that cybersecurity professionals can access their devices and their data. In this case, the biometric template can also be altered, deleted, corrupted, or accessed so that the user believes the sensor is damaged. Therefore, the user would disable the biometric protection of the device that is often used to encrypt data stored on the device. In addition to this, the unauthorized user may download and store the biometric template for future usage in cybercrimes. Hackers, through access to this data, may sell it to other criminals or use it for themselves to create fake biometrics (like gummy fingerprints) for gaining access to bank accounts, other financial data, or personally identifiable information (PII).

Smartphones and similar devices can be connected together using USB cords. These USB cords may often be used for charging the phone. However, cybercriminals could connect to unauthorized smartphones using USB cords for accessing and exploiting the data on the phone. In the Android operating system, a USB cord connection does not require password authentication to obtain access to the device. The biometric template stored on the device could be altered, damaged, changed, or deleted through access with a USB cord. Also, the biometric template can be downloaded and stolen for future usage in accessing the device. In most cases, the cybercriminal will sell the information on the dark web for financial gain.

Cybersecurity professionals can help smartphone users protect their biometric data in several ways. The easiest way is to provide free and quick education to owners of smart devices to safeguard against human error. This education can take the form of micro credentials that can be provided through social networking sites like LinkedIn and Facebook. These micro credentials could positively contribute towards the smartphone owners' career and may help to protect their data in the case of human error. Additionally, cybersecurity professionals must work to help owners easily establish applications that should have access to very sensitive data and those that do not require it. Owners should be easily able to whitelist or create secure sandboxes where personally identifiable data like biometric templates can be stored on the device securely. Finally, it is necessary that cybersecurity professionals lobby Congress and the United States to help create policy related to smartphone application data usage, access, and transparency. Children and adults with limited cognitive ability for understanding application disclosures are not prevented from downloading and installing applications that wish to exploit their personally identifiable data. Countries like the United Kingdom and Australia are working to ensure that its citizens' privacy is protected against technologies and the companies that create them. Therefore, cybersecurity lobbyists are needed to help create ethical, transparent, and accountable policy surrounding smartphone applications.

Biometric templates stored on smart devices should also be encrypted and alterable via two-factor authentication. In the case of connection hardware, cybersecurity professionals could work with the operating systems to require a password in all cases of connection to the device using USB cords.

Although most people are familiar with using biometric sensors to access their smartphones, more people interact regularly with biometric hardware technology in their doctor's office. Thermometers, blood pressure cuffs, blood oxygen oximeters, and respiration cuffs are used by nurses and doctors to assess the overall health of patients during hospital stays and visits. Data gathered from these devices include heart rate variability, blood pressure, skin temperature, galvanic skin response, and lung oxygen capacity. This information, the potential threats, and opportunities to secure healthcare biometrics are described more in the next section.

3.2 Healthcare Biometrics

Healthcare is one of the first industries to equip medical sensor technology with the ability to access and communicate data across networks. Blood pressure, heart rate, brainwave, and blood oxygen oximeter are all devices embedded with WiFi to communicate data to a centralized data location. Before fitness trackers and smart watches, healthcare professionals had the ability to collect data from a sensor and download it to a central location using USB connections and removable memory. However, now with WiFi technology, no longer do healthcare professionals have to connect the device to move the data securely to a central storage, they can now upload the data directly or automatically create processes that upload the data to various locations without much thought. Signals like heart rate variability, blood pressure, respiration, and temperature are now stored, in most cases, in cloud-based locations. Cloud storage locations are Internet-based memory storage locations. Medically-implanted healthcare sensors like pacemakers, smart hips, and the newest brain-based sensor called the Neuralink also connect and upload data to the cloud. Signals like beats per minute, heart rate, and heart rate variability are communicated to the cloud via a pacemaker. Signals associated with walking including gait, steps per minute, and the angle of the smart hip are moved to the cloud daily. Also, included in medically-implanted sensors are the name, contact information, and manufacturing information associated with the sensor. This data is communicated to the cloud with every upload. Doctors and patients can view the data by logging into a healthcare portal via the Internet. In some cases, these portals are accessible only through two-factor authentication using both a password and a secure code that is sent to a smartphone after the user enters the correct passcode. However, in most cases these healthcare sensors do not employ this technology.

With both areas of biometrics, a vast amount of data is usually communicated between the device and the network used to store the data, opening the door to good and bad exploitation.

4 EMOTIONAL AI OR AFFECTIVE COMPUTING

Emotional AI is an area of AI concerned with giving computers the ability to see, read, listen, and respond to human emotional information using faces, gestures, behaviors, and other physiological information (Picard, 2006). There are several ways that emotional AI tools can gather human physiological data, including (ChatGPT, 2023):

1. **Facial expression analysis**: One of the most common methods used by emotional AI tools is to analyze facial expressions to detect emotions. This is done by using computer vision algorithms to track and analyze changes in facial expressions, such as the movement of the mouth, eyebrows, and eyes.
2. **Voice analysis**: Emotional AI tools can also analyze the tone, pitch, and other characteristics of a person's voice to detect emotions. This is done by using natural language processing (NLP) algorithms to analyze the words and sounds used in speech.
3. **Galvanic skin response**: Some emotional AI tools use sensors to measure changes in a person's skin conductance, which can indicate emotional arousal. This is done by measuring the electrical conductivity of the skin, which changes when a person experiences emotional arousal.
4. **Heart rate variability**: Another method used by emotional AI tools is to measure a person's heart rate variability, which can indicate emotional arousal and stress levels. This is done by using sensors to measure the variation in time between heartbeats, which can be used to detect changes in emotional state.
5. **Brain activity**: Some advanced emotional AI tools use electroencephalography (EEG) to measure brain activity, which can provide a more direct measurement of emotional state. This is done by using sensors to measure the electrical activity in the brain, which can indicate the presence of specific emotions.

Data gathered through affective computing techniques are usually highly sensitive. Like biometrics, this data is sent over networks that can be exploited. Additionally, the data repositories that store the data must be secured properly through encryption techniques if left in the cloud. Most systems now have cloud-based architectures that tool developers and users forget can also be exploited. When using cloud-based databases, a private cloud that is secured against outside intrusions like adversarial attacks or brute force attacks should be considered. Adversarial attacks can trick emotional AI systems into misidentifying emotions or recognizing fake emotional expressions, leading to incorrect or dangerous responses to the users.

As emotional AI techniques that exploit heart rate, brain signals (Shakshi & Jaswal, 2016), or other physiological data lend themselves towards unique ways to establish passwords or provide two-factor authentication, they may be susceptible to brute force attacks. Brute force attacks are ways hackers try to manually or automatically try to crack a password or encryption key by systematically trying every possible combination until the correct one is found. Automated brute force techniques, in theory, could try millions of number combinations until they match the unique patterns found in a user's heart rate or heart rate variability. Similarly, Functional Magnetic Resonance Imaging (FMRI) or brain wave signals can be used as a user's private key to unlock their encrypted data. Normally, EEG signals are susceptible to fatigue in humans, but it is possible for a baseline unique signal to be created for a particular human that adjusts for abnormal conditions like time of day and health conditions affecting the brain. Additionally, the brain wave signal could be transformed to a more unique sequence using a fast Fourier transform to account for the power of the signal that may be unaffected by fatigue in the human. In this case, if a hacker knows that the private key number sequence is a fast Fourier transform over the original EEG or FMRI signal, they can easily access a person's private data in a short amount of time through automated brute force software.

Computers can passively gather human physiological data for use in emotional AI in several ways including (ChatGPT, 2023):

1. **Webcam monitoring**: Some emotional AI tools use computer vision algorithms to passively monitor a person's facial expressions through their webcam. The algorithms analyze changes in facial expressions, such as the movement of the mouth, eyebrows, and eyes, to detect emotions.
2. **Microphone monitoring**: Computers can also passively gather human physiological data through microphone monitoring. Emotional AI tools can analyze a person's voice, tone, pitch, and other characteristics to detect emotions.
3. **Wearable sensors**: Computers can also passively gather human physiological data through wearable sensors, such as smartwatches or fitness trackers. These sensors can measure changes in heart rate, skin conductance, and other physiological markers, which can be used to detect changes in emotional state.
4. **Eye-tracking**: Some emotional AI tools use eye-tracking technology to passively gather physiological data. By tracking a person's eye movements, the algorithms can detect changes in visual attention, which can be an indicator of emotional arousal.

With passive gathering of data, humans may forget they are being monitored, thereby opening them up for a myriad of attacks in addition to those already discussed. An attacker can feed the emotional AI system with carefully crafted inputs designed to mislead the system and produce incorrect results. For example, a hacker can provide the system with an image of a facial expression that does not match the actual emotion of the person, leading the system to misidentify the emotion. Also, an attacker can slightly modify the facial expression in an image to evade detection by the system. Furthermore, the machine learning model that drives the AI may become poisoned if the model is updated through reinforcement learning using malicious data designed to misclassify emotions or produce biased results.

Humans are often the weakest link in HCAI. Emotional AI tools that use voice or text-based interactions can be manipulated by social engineering attacks, where an attacker can trick a user into revealing sensitive information or taking unwanted actions if they think they are interacting with the AI and not a third party. In these cases, more training will need to be provided to humans on spotting chat-based AI systems that can operate through audio or text-based interactions. Systems like ChatGPT can mimic the empathy expressed by humans in conversations. Humans need to properly test the functionality of systems like ChatGPT and others so they can spot them. Behavioral cybersecurity experts will need to develop software training modules that help humans spot the difference in the communication styles of humans and these AI chatbots. Finally, interdisciplinary researchers from psychology, computer science, and sociology will need to come together on ethical guidelines surrounding potential uses of chat-like systems for social engineering. Along with regular security audits and anomaly detection audits to guard against attacks in emotional AI, more studies will reveal the true unattended consequences of hacking these emotionally-aware and controversial systems (Noone, 2022).

5 PERSONALITY COMPUTING AND PERSUASIVE AI

Personality computing is a field of study that focuses on using technology and data to understand and measure human personality traits. Personality traits are stable patterns of thoughts, feelings, and behaviors that distinguish individuals from one another. Personality computing typically involves using data from a variety of sources, including social media activity, text analysis, and physiological data, to create models that can predict and analyze personality traits. These models can be used in a variety of applications, such as hiring and recruitment, personalized marketing, and mental health assessment. Adversarial attacks can be used to manipulate the results of personality computing algorithms. An example is adding small changes to a person's online activity that may predict negative traits like narcissistic personality disorder or antisocial personality disorder. Also, personality computing algorithms can be used in chatting software to appear more human. An attacker can cause a customer-facing chat software to appear mean, unhelpful, and antisocial thereby causing the reputation and bottom line of the company to suffer.

Persuasive AI refers to the use of AI technology to persuade or influence people's beliefs, attitudes, or behaviors. This can be achieved through various means, such as personalized recommendations, targeted advertising, and even political campaigns. Persuasive AI can be used for both positive and negative purposes. On the positive side, it can be used to promote healthier lifestyles, encourage eco-friendly behavior, or support social causes. On the negative side, it can be used to spread misinformation, manipulate public opinion, or exploit vulnerable individuals. Hackers or persons that want to promote disinformation or "fake news" can use persuasive AI to produce large amounts of content that contains falsehoods, misinterpreted statistical data, and other negative opinions about a political, cultural, or social group. Persuasive AI used in decision-making tasks can be manipulated into making biased or unethical decisions, such as denying loans or job applications based on protected characteristics like race, gender, sex, or sexual orientation.

6 DISCUSSION ON OPPORTUNITIES FOR NEW APPROACHES TO CYBERSECURITY FOR HUMAN-CENTERED AI

Much of the research and innovations surrounding cybersecurity relate to network security and physical security of a computing device. Little research has examined the difficulty in creating cybersecurity innovations for AI but even less study has been devoted to cybersecurity approaches for HCAI. This section discusses the approach, talks about the challenges surrounding it, and other opportunities that need to be explored to fully realize the innovation.

Previously described in the above sections were ways hackers could attack HCAI technology. These attacks pertained to misleading or biased input, malicious data poisoning the machine learning model predictions, identifying weaknesses in the AI, and fooling the user or system into thinking interactions are with a valid human user. To overcome these will require thinking outside of the normal cybersecurity toolkit because like humans these systems can mimic humans very well. (Think of ChatGPT and its current and future uses.) There is an opportunity to start to employ large language models or NLP techniques that can examine, learn, and identify the output format of ChatGPT-like systems that are used by hackers to mimic humans and gain unauthorized access to data. Biometric or HCAI systems that utilize human physical or physiological data need ways to verify that the data is gathered from a live human. Current biometric techniques for liveness detection include use of fast Fourier transforms applied to the iris, heart rate variability, and EEG signals. This technique has also been shown to work on images that can be used in facial expression recognition software.

Usable security techniques employ CAPTCHAS (Moradi & Keyvanpour, 2015) as a simple technique to indicate a person is a human and not a bot. HCAI techniques will have to develop more interactive CAPTCHAS that include multimodal information for accessing if a person is human or not. Deepfake technologies have been used to create human voices and faces. However, interactive CAPTCHAS that randomly combine an image task with a word task may be an opportunity for exploration with use in HCAI software.

Finally, social engineering is a problem as humans want to interact and connect with other humans. An opportunity to explore the creation of models that mimic true, deep human emotion needs to be studied. Empathy, happiness, and excitement have been mimicked by chatbot software. But often humans can spot the genuine emotions that are displayed by real humans to other humans. Study of the features, structures, and attributes that indicate true genuine emotion in humans needs to be researched. A crowdsourced dataset could easily be created from fake ChatGPT software and genuine chatting data between friends, family, and loved ones. Along with this to ensure that this dataset is not used for learning by AI, ethical, transparent, fair, and accountable guidelines need to be developed along the usage of the dataset. Also, only non-law enforcement agencies would have access to this dataset.

7 CONCLUSION

This section explored threats and opportunities for cybersecurity approaches designed for HCAI, specifically emotional AI and biometrics. HCAI is an advancing field and the uniqueness of the cybersecurity flaws that present themselves with attacks of this kind of software are numerous. Furthermore, HCAI technology is designed to either gather data from the human or interpret and sense data about the human's emotion, behavior, and/or identity. This is tricky and the sensitive nature of this human-centered data makes it particularly important that new innovations be developed or researched with the human and tool's purpose in mind. Innovations that help cybersecurity researchers utilize their go-to techniques in the toolbox with modifications may help to combat the privacy and security issues that may arise with these sensitive systems (McStay, 2020). Specifically, this area is an ever-expanding area that needs interdisciplinary voices to combat the issues that may arise with each new tool creation. This will require that psychologists, sociologists, computer scientists, and engineers allow each other to speak and contribute to the studies needed for understanding and examining the flaws that may arise. As the uses of these tools grow to new sectors including healthcare, business hiring decisions, decision support, and human mimicking, the voices welcomed in the space need to be diverse and varied.

REFERENCES

1. ChatGPT, personal communication, February 11, 2023.
2. Li, Y., & Hilliges, O. (Eds.). (2021). *Artificial Intelligence for Human Computer Interaction: A Modern Approach.* pp. 463–493. Cham: Springer.
3. McStay, A. (2020). Emotional AI, soft biometrics and the surveillance of emotional life: An unusual consensus on privacy. *Big Data & Society*, 7(1), 2053951720904386.
4. Moradi, M., & Keyvanpour, M. (2015). CAPTCHA and its alternatives: A Review. *Security and Communication Networks*, 8(12), pp. 2135–2156.
5. Noone, G. (2022). Emotion Recognition is Mostly Eneffective. Why are Companies Still Investing in It?, *Tech Monitor*. Available at: https://techmonitor.ai/technology/emerging-technology/emotion-recognition (Accessed: February 24, 2023).
6. Picard, R. W. (2000). *Affective computing*. MIT press.
7. Shakshi, R. J., & Jaswal, R. (2016). Brain wave classification and feature extraction of EEG signal by using FFT on lab view. *Int. Res. J. Eng. Technol*, 3, pp. 1208–1212.

Index

Printed in the United States
by Baker & Taylor Publisher Services